T0251278

ENGINEERING TECHNOLOGY PROBLEM SOLVING

ENGINEERING TECHNOLOGY PROBLEM SOLVING

Techniques Using Electronic Calculators

Second Edition

Houston N. Irvine

Hawkeye Institute of Technology
Waterloo, Iowa

CRC Press
Taylor & Francis Group
Boca Raton London New York

CRC Press is an imprint of the
Taylor & Francis Group, an **informa** business

Library of Congress Cataloging-in-Publication Data

Irvine, Houston N.
  Engineering technology problem solving: techniques using
electronic calculators / Houston N. Irvine. -- 2nd ed.
    p. cm.
  Includes bibliographical references and index.
  ISBN 0-8247-8606-8 (acid-free paper)
  1. Engineering mathematics. 2. Engineering--Data processing.
I. Title.
TA331.I78 1992
620'.001'51028--dc20   91-42427
                                                    CIP

Copyright © 1992 by MARCEL DEKKER   All Rights Reserved

Neither this book nor any part may be reproduced or transmitted in any
form or by any means, electronic or mechanical, including photocopying,
microfilming, and recording, or by any information storage and retrieval
system, without permission in writing from the publisher.

MARCEL DEKKER
270 Madison Avenue, New York, New York, 10016

Preface

As with the previous edition, this text is designed for use by beginning engineering and technical students and as a handbook for calculator applications. When the first edition of this book was published in 1981, it dealt with the two most popular calculators at that time, the Texas Instruments TI-55 and the Hewlett-Packard HP-33E. Since then quite a few changes and improvements have been made in calculators. Memories are now labeled with letters rather than numbers, which makes it possible to store data in memories that correspond to symbols in equations. The keyboards have been changed to accommodate the use of additional functions. And, in some cases, the order of entry of data has been changed.

This edition deals with three makes of calculators: Casio, Texas Instruments, and Hewlett-Packard. It should be noted that each of these manufacturers produces models other than the ones covered in this text. We have tried to choose the models that are most likely to be used by students. The routines for sample problems have been rewritten to suit the current models of these calculators. Because each brand of calculator offers some unique functions, the last three chapters describe special functions and programming techniques of each.

We wish to acknowledge permission for the use of material from the instruction manuals for their calculators from the manufacturers Texas Instruments, Casio Inc., and Hewlett-Packard.

Houston N. Irvine

Contents

ENGINEERING
TECHNOLOGY
PROBLEM
SOLVING

1 · Introduction to Problem Solving

I. Problem-Solving Format

The wonders of modern civilization would not have been possible without the efforts of the engineering profession. Designing and creating the structures and machinery for modern living involves a great deal of problem solving. Engineers must be able to design equipment which is safe and reliable. They must also be able to design for cost effectiveness so that their projects are affordable and competitive. The technician, as an engineering assistant, is responsible for much of the problem solving involved.

The solution of technical problems requires developing habits of neatness and organization. While this may at first seem tedious, it should soon become apparent that the extra effort is well justified. All problem-solving work is to be lettered. Special care should be taken to ensure that all figures are formed accurately, so that there is no chance of their being misread. Lettering should be done with a well-sharpened pencil, preferably H or 2H. When erasures are necessary, the student should use a clean drafting eraser. Erasures should be clean. No smudges or smears are tolerated. All decimal points must be distinct.

Where geometrical relationships are involved, a carefully drawn sketch should be provided, with all pertinent dimensions and other information clearly shown. As more experience is gained in this procedure, the student will find that a well-prepared sketch is a valuable aid in the solution of a problem.

Next is the statement of the problem. What information is given and what is required? If a formula is involved, state the formula, followed by the substitution of the given data in the formula. Make sure that the proper units are used. When the problem has been solved, underline the answer so that it stands out clearly. Make sure that the proper units are part of the answer. A sample problem solution sheet is shown in Fig. 1.1.

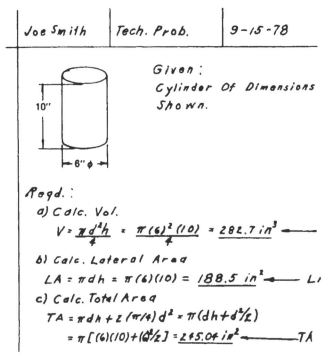

FIG. 1.1 Sample problem solution sheet.

Observe and follow this format carefully. Among the reasons for taking this case in setting up technical problems are the following:

1. Accuracy. Mistakes can be expensive.
2. Ease in interpretation. In technical and engineering work, other people will have occasion to check your calculations from time to time.
3. As problems become more lengthy and complex, you can easily lose track of what you are doing if your problem solution is not well organized.

Your problem solution should be clear enough that you can look at it a year from now and easily understand what you did.

Before learning to depend on the calculator for answers, you should train yourself to approximate answers by mental arithmetic. Improper entries of numbers can lead to answers that are greatly in error. If the calculator answer agrees fairly closely with the approximate answer, you can be confident you have solved the problem correctly. To explain methods of approximation we will start with fairly small numbers. Consider the calculation of volume for the sample problem:

$$v = \frac{\pi d^2 h}{4} = \frac{\pi (6)^2 (10)}{4}$$

We know that the value of π is slightly more than 3. Approximating:

$$v = \frac{(3)(6)^2(10)}{4} = 270$$

As we can see, this is fairly close to the correct answer. A general rule to follow is to round off all numbers to whole numbers of one or two digits and apply them in the equation. When the figure beyond the last place to be retained is less than 5, the figure in the last place retained is left unchanged. When the figure beyond the last place to be retained is greater than 5, the figure in the last place retained is increased by 1. Following are some examples of calculator solutions compared with approximate solutions:

EXAMPLE 1. $\dfrac{6.5 \times 7.3}{2.9} = 16.362.$

Approximating: $\dfrac{7 \times 7}{3} = 16.3$

EXAMPLE 2. $\dfrac{(5.65)(3.75 + 9.05)}{17.6} = 4.109$

or approximately,

$\dfrac{(5.7)(3.7 + 9)}{18} \cong \dfrac{(6)(13)}{18} \cong \dfrac{78}{18} \cong 4.3$

EXAMPLE 3. $\dfrac{(15.6 + 3.8)}{(5.3 + 2.9)} = 2.366$

or approximately,

$\dfrac{(16 + 4)}{(5 + 3)} = \dfrac{20}{8} = 2.5$

EXAMPLE 4. $\dfrac{(17.9 + 5.3)(4.7 + 1.2)}{2.9} = 47.2$

or approximately,

$\dfrac{(18 + 5)(5 + 1)}{3} = \dfrac{(23)(6)}{3}^{2} = 46$

Methods for dealing with more complex numbers will be taken up in Section 1.6.

II. Basic Functions and Operations

The functions described below are common to most calculators of the scientific type. In this text we are describing operating procedures for three of the most popular makes of scientific calculators. The makes and models of the calculators described in this text are: the Texas Instruments TI-68, the Casio FX-7000GA, and the Hewlett Packard 32SII. There are different models of each make, but the operating procedures for each are similar. These calculators contain a number of special functions in addition to the standard ones mentioned below. They deal with such subjects as statistics, programming, integration, simultaneous equations, and graphics. Because of the number of functions covered in the limited space available, it is necessary to place more than one function on a key.

On the TI-68 there are two functions shown on each key. The lower one, shown in black, requires no shift. The upper one, shown in yellow, is executed when the "2nd" key is pressed. Above each key, shown in blue, is a third function. It is executed when the "3rd" key is pressed. The inverse of a function is executed when the "INV" key is pressed. The TI-68 uses the symbol "X^{-1}" for reciprocals and the symbol "$(-)$" for the change sign function.

On the Casio FX-7000GA the function appearing above the key is executed when the "SHIFT" key is pressed. "X^{-1}" is used for the reciprocal function. "$(-)$" is used for the change sign function. "EXE" is used in place of "$=$." The "EXP" key is used to enter scientific notation. Other special symbols appearing on the keyboard will be explained as problems involving their use are discussed.

On the Hewlett-Packard 32SII the function colored orange appearing above the key is executed when the ⏎ key is pressed. The function appearing in blue is executed when the Γ key is pressed. There is a third set of functions available when the menu keys are pressed. (See the manual.) For example, when the "⏎ DISP" key is pressed a menu appears:

FX SC EN ALL

If one wishes to fix the decimal point, he should press the key in the row below FX and specify the number of decimal places desired. Other menu keys will be explained as problems involving their use are discussed.

All three of the calculators being discussed have a constant memory. Any data in memory registers or programs will be retained when the calculator is switched off.

When an incorrect number is entered on the TI-68 it may be erased by pressing the "CLEAR" key. If a single digit is incorrect, press the "←" key to the position of the incorrect digit and overwrite with the correct one. To clear the entire memory, press "3rd RESET." The display shows "CLR MEM ? YN." If you are sure there is nothing you want to save, press Y.

For the Casio FX-7000GA an incorrect entry may be cleared by pressing the "AC" key. If a single digit is incorrect, press "←" to the position of the incorrect digit, press the "DEL" key and enter the correct digit. To clear the entire memory, press "SHIFT CL."

The HP-32SII retains the last number appearing in the display when the calculator is switched off. It is available for subsequent calculations when the calculator is switched on again. If one wants to clear the display, press the "C" key. If a single digit is incorrect, press the "←"

key to its position and overwrite. If " ⌐ CLEAR" is pressed a menu appears:

VARS   ALL   Σ

If one wishes to clear the entire memory, press the key below ALL. A prompt, "CLR ALL ? YN" appears. Press Y to clear.

The following is a description of the basic functions you will find on the keyboard:

1. The basic four functions, addition, subtraction, multiplication, and division.
2. The reciprocal function $1/x$
3. The square root function $\sqrt{x}$
4. The x^2 function
5. The Y^x function, which allows you to raise a number to any power
6. The $x\sqrt{y}$ function, which enables you to take any root of a number
7. The trigonometry functions, sin, cos, tan
8. The common logarithm function, log x
9. The natural logarithm function, ln x
10. The scientific notation key, <u>EE</u>
11. The storage key <u>STO</u>, which enables you to store a number
12. The recall key <u>RCL</u>, which enables you to recall a number from storage
13. The change sign key $+ / -$
14. The π key, which enters the value of π directly
15. The exchange key x ⤲ y, which enables you to exchange registers

In this book we progress a step at a time from the simpler to the more complex functions. Practical applications in problem solving are presented as each function is explained.

III. Algebraic and Reverse Polish Logic

Pocket calculators fall into two categories: those using algebraic logic and those using Lukasciewicz (reverse Polish) logic. Texas instruments and Casio calculators use algebraic logic. These calculators have an " = "

or "EXE" key on the keyboard. Entry of a problem is in regular algebraic order:

$2 + 3 = 5$

$6 - 4 = 2$

$5 \times 4 = 20$

$55 \div 11 = 5$

Calculators using reverse Polish logic have an "ENTER" key, but no "=" key. At first this may seem awkward, but as you learn to work with more complex problems, certain advantages become apparent. The HP-32SII uses this type of logic.

Other differences between the two types of logic will be explained as we take up functions other than the basic ones of addition, subtraction, multiplication, and division.

IV. Hierarchy of Operations: Algebraic Logic

Calculator hierarchy determines the order of completion of each calculator function. When functions are used individually, hierarchy is of little consequence. However, when functions are used collectively in the solution of an algebraic equation, the order of completion is important. The complete list of priorities for algebraic hierarchy follows (Texas Instruments, 1977a).

1. Special functions (trigonometric, logarithmic, square, square root, factorial, e^x, 10^x, percent reciprocal, and conversions) immediately replace the displayed value with its functional value.
2. Percent change ($\Delta\%$) has only the ability to complete other percent change operations.
3. Exponentiation (Y^x) and roots ($x\sqrt{y}$) are performed as soon as special functions and percent change are completed.
4. Multiplication and division are performed after the above operations and other multiplication and division are completed.
5. Addition and subtraction are performed only after completing all operations through multiplication and division as well as other addition and subtraction.
6. Equals completes all operations.

The solution of problems involving the four basic functions has already been explained for both types of logic. One of the most convenient features of the calculator is its ability to handle problems involving chain multiplication and division. It is only necessary to enter the numbers in succession:

$$\frac{7.5 \times 83}{0.57} = 1092.1052$$

For algebraic logic the procedure is as follows:

ENTER	PRESS	DISPLAY
7.5	×	
83	÷	
0.57	=	1092.1052

For reverse Polish logic the procedure is as follows:

PRESS	DISPLAY
7.5 ENTER	
83 ×	622.5
0.57 ÷	1092.105

For the algebraic operating system parentheses are used to enter complex equations in straightforward order. If an expression is enclosed in parentheses, it is evaluated without pressing the = sign. For example, press (5 × 9) and 45 will be displayed. Parentheses can be used in this manner to enter more complex expressions.

EXAMPLE 5. $\dfrac{4 \times (5 + 9)}{(7 - 4)} = 18.667$

For algebraic logic, using parentheses:

ENTER	PRESS	DISPLAY
4	× (	4
5	+	5
9	)	14
	÷	56
	(	
7	−	7
4	)	3
	=	18.667

Remember to close parentheses at the proper places.

For reverse Polish notation the procedure is simpler and uses fewer strokes:

PRESS	DISPLAY
4 ENTER	
5 ENTER	
9 + ×	56
7 ENTER	
4 − ÷	18.667

A sum of products is handled as follows:

EXAMPLE 6. $(5 \times 6) + (7 \times 8) = 86$

Calculator routine for algebraic logic:

ENTER	PRESS	DISPLAY
5	×	
6	+	
7	×	
8	=	86

Calculator routine for reverse Polish notation:

PRESS	DISPLAY
5 ENTER	
6 ×	30
7 ENTER	
8 ×	56
+	86

V. Rounding Off

Most calculator displays are capable of showing answers to at least eight-place accuracy. However, for the majority of calculations such accuracy is meaningless. First, no calculation is more accurate than the least accurate data entering into it. In most practical cases measurements are not likely to be more accurate than four significant figures and are often even less accurate. Second, it is wasteful of time and effort to record eight-digit numbers. For general practice we recommend rounding off answers to four significant figures unelss you have been otherwise instructed.

For proper rounding off, we offer the following rules established by the American Standards Association:

1. When the figure beyond the last place to be recorded is less than 5, the figure in the last place retained is unchanged.
2. When the figure beyond the last place to be recorded is greater than 5, the figure in the last place retained is increased by 1.

3. When the figure beyond the last place to be recorded is 5 and
   a. The figure in the last place retained is odd, it should be increased
      by 1.
      Example: 17435 rounds off to 17440,
   b. The figure in the last place retained is even, it should be kept
      unchanged.
      Example: 3545 rounds off to 3540.

EXAMPLE 7.   Round off the number 18635 to 4 places, 3 places, and
2 places.

Solution:

4 places	18640
3 places	18600
2 places	19000

For answers less than 1, answers to four significant figures mean the
first four digits following zero or a series of zeros. For example, the
number 0.003564 would not be rounded off.

The models of calculators being considered all have routines for automatically rounding off to any desired accuracy to the right of the
decimal point. For the TI-68 the "2nd FIX" key is used. For four digit
accuracy to the right of the decimal point press "2nd FIX 4." The number
of digits displayed may be changed by again pressing the "2nd FIX"
followed by the number of digits desired. The amount carried in the
memory and applicable to subsequent calculations is accurate to ten
places regardless of the number of places shown in the display.

For the Casio FX-7000GA the number of decimal places is set by
pressing the "MODE 7" keys followed by the number of places desired,
followed by "EXE." To set four places, press "MODE 7 4 EXE."

For the HP-32SII the number of decimal places is set by the following
procedure: press " ⌐ DISP." A menu appears:

FX SC EN ALL

Press the key in the row immediately below "FX" and enter the number
of decimal places desired.

Now is the time to start practicing the use of your calculator. Work
Exercise 1.1 on problem paper, using the recommended format.

VI. Scientific Notation

There are many engineering and scientific calculations that involve both very large and very small numbers. For example, the elastic modulus of steel is 30,000,000 psi, while the coefficient of linear expansion for steel is 0.0000065 in/in. To deal with such numbers, it is convenient to use scientific notation. Scientific notation consists essentially of expressing a number as a power of 10. For example, 3000 is expressed in scientific notation as 3×10^3, or 3E3. The number 0.0000065 is expressed in scientific notation as 6.5×10^{-6}, or 6.5E-6. It is recommended that you use the latter notation when writing numbers in scientific notation.

Your calculator will handle numbers in scientific notation from 9.9E-499 to 9.9E499. For entering numbers in scientific notation use the key marked "EE" on the TI-68, the key marked "EXP" on the Casio FX-7000GA, and the key marked "E" on the Hewlett Packard HP-32SII.

Calculator routine for the TI-68:

ENTER	DISPLAY
1.2 EE 5	1.2 E05

Calculator routine for the FX-7000GA:

ENTER	DISPLAY
1.2 EXP5	1.2 E05

Calculator routine for the HP-32SII:

ENTER	DISPLAY
1.2E5	1.2E5

For entering negative powers of ten, the change sign key is pressed before the power is entered.

EXAMPLE 8. Enter 6.5 E-6.

Calculator routine for the TI-68:

PRESS	DISPLAY
6.5 EE(−)6	6.5 E-6

Calculator routine for the FX-7000GA:

PRESS	DISPLAY
6.5 EXP (–)6	6.5 E-6

Calculator routine for the HP-32SII:

PRESS	DISPLAY
6.5 E +/ –	6.5 E-6

EXAMPLE 9.   Perform the following operation using scientific notation:
$$\frac{3E7 \times 2.5E4}{3.6E4} = 2.083E7$$

Calculator routine for the algebraic notation:

ENTER	PRESS	DISPLAY
3	EE	
7	×	3 07
2.5	EE	
4	÷	7.5 11
3.6	EE	
4	=	2.083 07

Calculator routine for reverse Polish notation:

PRESS	DISPLAY
3 E7 ENTER	
2.5 E4 ×	
3.6 E4	2.083 E7

On all of these calculators the result of operations on numbers in scientific notation may not appear in the display in scientific notation unless the display is set in the scientific notation mode.

For the TI-68 this is accomplished by pressing "3rd ScEn" twice.

For the FX-7000GA press "MODE 8" followed by the number of decimal places followed by "EXE".

For the HP-32SII, press "¬ DISP". The menu appears:

FX SC EN ALL

Press the key in the row below "SC" and enter the number of decimal places desired.

A number in the display in fixed point notation may be changed to scientific notation if the calculator is already set in the scientific notation mode.

EXAMPLE 10. Change the number 30,256 to scientific notation rounded off to four places.

Calculator routine for the TI-68:

PRESS	DISPLAY
30,256 =	3.0256 E04

Calculator routine for the FX-7000GA:

PRESS	DISPLAY
30,256 <u>EXE</u>	3.0256 E + 04

Calculator routine for the HP-32SII:

PRESS	DISPLAY
30,256 <u>DISP (SC)</u>	3.0256 E4

At this point let us also take up the operation of extracting the square root of a number. This operation is essential to the solution of many scientific problems. To obtain the square root of a number, it is only necessary to enter the number and press the $\sqrt{x}$ key. For calculators that have an x^2 key, the procedure for squaring a number is to enter the number and press the x^2 key.

EXAMPLE 11. $\dfrac{\sqrt{3.56E6} \times (4.5E3)^2}{710} = 5.453E7$

Calculator routine for algebraic notation:

PRESS	DISPLAY
$\sqrt{}$ 3.56 EE6 ×	
4.5 EE3 x^2	
÷ 710 =	5.38 E07

Calculator routine for reverse Polish notation:

PRESS	DISPLAY
3.56 E6 $\sqrt{x}$	
4.53 E3 ⌐ x^2 ×	
710 ÷	5.38E7

The use of powers of 10 in scientific notation gives us a key to approximating answers to problems involving both large and small numbers. It can be seen that by using scientific notation, any number can be expressed as one or two digits multiplied by a power of 10. Now you can recall from basic algebra that when two numbers having the same base and an exponent are multiplied, the exponent of the product is equal to the sum of the exponents of the numbers being multiplied:

$$(x^2)(x^3) = x^5$$

In a similar manner, a division involves subtracting the exponent of the divisor from the exponent of the dividend:

$$\frac{x^5}{x^2} = x^3$$

To raise a number to a power, multiply the exponent by the power:

$$(x^2)^3 = x^6$$

To take a root, divide the exponent by the root:

$$\sqrt{x^6} = x^3$$

Use the following two rules for approximating calculations involving large numbers:

1. Change the numbers to scientific notation and round off to one or two places.

2. For taking square roots of numbers that are odd powers of 10, it is necessary to change the number under the radical to even powers of 10. For example, 3.5E5 could be approximated as 36E4.

Now let us try approximating some problems involving larger numbers.

EXAMPLE 12. $\dfrac{3560 \times 824}{390} = 7521.6$

Solution: Approximating,

$$\dfrac{4E3 \times 8E2}{4E2} = 8E3 \text{ or } 8000$$

EXAMPLE 13. $\dfrac{4.55 \times 584 \times 376}{215 \times 3E7 \times 540} = 2.868\text{E-}7$

Solution: Approximating,

$$\dfrac{4.5 \times 6E2 \times 4E2}{2E2 \times 3E7 \times 5.5E2} = \dfrac{108}{3.3E8} = 3.3\text{E-}7$$

EXAMPLE 14. $\dfrac{175 \times 370}{0.006} = 1.079\text{E}7$

Solution: Approximating,

$$\dfrac{2E2 \times 4E2}{6\text{E-}3} = \dfrac{8E4}{6\text{E-}3} = 1.3\text{E}7$$

EXAMPLE 15. $\dfrac{\sqrt{3.5E3} \times (3.9E5)^2}{590} = 1.525\text{E}6$

Solution: Approximating,

$$\dfrac{\sqrt{36E2} \times (4E3)^2}{6E2} = \dfrac{60 \times 16E6}{600} = 1.6\text{E}6$$

VII. Memory

All of the calculators being considered have twenty-six memory registers, numbered A through Z. This makes it convenient to identify memories with symbols in an equation. The methods of storing are slightly different for all three calculators.

To store a number in the TI-68, press the number then "STO" and "Alpha," followed by the name of the variable, followed by " =." A variable name must begin with a letter and may consist of one, two, or three characters. To store 25 in memory A the sequence is "25 STO Alpha A =." The number stored may be applied in an equation or may be recalled by pressing "RCL."

The procedure for the Casio FX-7000GA is similar except that the "→" symbol is used in place of "STO" and "EXE" is used in place of " =."

For the HP-32SII the procedure is simpler. For example, press "25 STO A." For application of the variable in an equation, press "RCL A."

EXAMPLE 16. Solve the equation $X^2 + 5X + 20$ for $X = 25.1$.

Calculator routine for algebraic notation:

PRESS	DISPLAY
25.1 STO ALPHA X =	
X^2 + 5x ALPHA X − 20 =	735.51

Calculator routine for reverse Polish notation:

PRESS	DISPLAY
25.1 STO A X^2	
5 ENTER RCL A x +	
20 −	735.51

EXAMPLE 17. Evaluate the equation $3X + 4Y − 5 XY$ for $X = 7.5$, $Y = 9.8$.

Calculator routine for algebraic notation:

PRESS	DISPLAY
7.5 <u>STO</u> <u>ALPHA</u> X =	
9.8 <u>STO</u> <u>ALPHA</u> Y =	
3x <u>ALPHA</u> X x^2 +	
4x <u>ALPHA</u> Y x^2 –	
5x <u>ALPHA</u> X x <u>ALPHA</u> Y =	185.41

Calculator routine for reverse Polish notation:

PRESS	DISPLAY
7.5 <u>STO</u> X	
9.8 <u>STO</u> Y	
3 <u>ENTER</u> <u>RCL</u> x x^2 x	
4 <u>ENTER</u>	
<u>RCL</u> Y x^2 x +	
5 <u>ENTER</u> <u>RCL</u> X x	
<u>RCL</u> Y x –	185.41

EXAMPLE 18. Evaluate $x^2 + 7$ for X = 9, X = 8.5 and total results.

Calculator routine for algebraic notation:

PRESS	DISPLAY
9 x^2 + 7 =	88
<u>STO</u> <u>ALPHA</u> A =	
8.5 x^2 + 7 =	79.25
<u>STO</u> <u>ALPHA</u> B =	
<u>ALPHA</u> A + <u>ALPHA</u> B =	167.25

Calculator routine for reverse Polish notation:

PRESS	DISPLAY
9 ⌐ x² 7 + STO A	88
8.5 ⌐ x² 7 +	79.25
RCL A +	167.25

It is frequently necessary to perform an operation on a series of numbers by a constant. The memory can conveniently be used for this.

EXAMPLE 19. Multiply the following series of numbers by the constant 7.56: 8, 10, 12.

Calculator routine for algebraic notation:

PRESS	DISPLAY
7.56 STO ALPHA A =	
8 × ALPHA A =	60.48
10 × ALPHA A =	75.56
12 × ALPHA A =	90.72

For the HP-32SII the " ⌐ LAST x" function can be used for operations with a constant.

Calculator routine for the HP-32SII:

PRESS	DISPLAY	COMMENT
8 ENTER	8	Enter first number
7.56 x	60.48	Multiply by constant
10 ⌐ LAST x	7.56	Retrieve constant
x	70.56	Multiply by constant
12 ⌐ LAST x	7.56	Retrieve constant
x	90.42	Multiply by constant

This chapter has covered the fundamentals of engineering problem-solving format and calculator operation. Diligent effort in working the exercises will help you to master the necessary skills.

Exercise 1.1

Solve by calculator and check by approximation:

1.1-1 $7.564 \times 3.15 \times 4.3 =$

1.1-2 $\dfrac{0.3125 \times 315}{0.6} =$

1.1-3 $\dfrac{7.5 \times 9.32}{8.15 \times 7.03} =$

1.1-4 $\dfrac{9.67 + 8.45}{3.09} =$

1.1-5 $(6 \times 150) + (7.5 \times 400) =$

1.1-6 $\dfrac{(4 \times 75) + (5.5 \times 150) + (8 \times 3.25)}{10.5} =$

1.1-7 $(0.5)(13.5 + 7.2) =$

1.1-8 $\dfrac{(9.37)(4.5 + 9.6)}{15} =$

1.1-9 $\dfrac{(13.7 + 9.65)}{(4.95 + 7.13)} =$

1.1-10 $\dfrac{(56.93 + 17.2)(4.3 + 1.55)}{3.045} =$

Exercise 1.2

Solve by calculator and check by approximation:

1.2-1 $30E6 \times 150 \times 6.56E\text{-}6 =$

1.2-2 $\dfrac{4.5E8 \times 350}{3.5E6} =$

1.2-3 $\dfrac{5 \times 17.65E3 \times 2.52E2}{384 \times 30E6 \times 227} =$

1.2-4 $\dfrac{5 \times 4.8E4 \times 2.88E2}{384 \times 30E6 \times 663} =$

1.2-5 $\dfrac{7}{6} [5.64E4 + (4 \times 26.4E4) + 62.5E4] =$

1.2-6 $\dfrac{18}{6} [1.84E5 + (4 \times 3.46E4) + 5.63E4] =$

1.2-7 $\dfrac{50 \times 10 \times 5 \times (625 - 100 - 25)}{6 \times 25 \times 30E6 \times 4.91E - 2} =$

1.2-8 $\dfrac{3.4E5}{5.6E4} + \dfrac{7.6E3}{8.9E5} =$

1.2-9 $\dfrac{(2.15E4 + 9.3E5)}{(4.2E6 + 7.1E4)} =$

1.2-10 $\dfrac{(\sqrt{3.19E6} + \sqrt{9.15E5}) \times 15.6}{(1.5E - 3^2 + 4.82E - 2^2)}$

Exercise 1.3

1.3-1 Evaluate the equation $x^2 + 3x + 17$ for $x = 105$.
1.3-2 Evaluate the equation $2.5x^2 + 7.3y^2 + 3.4xy$ for $x = 21.5$, $y = 17.4$.
1.3-3 Evaluate the equation $x^2 + 35.7$ for $x = 17.5$ and $x = 29.2$ and total the results.
1.3-4 Using the constant routine multiply the following numbers by the constant 0.73: 12, 83, 75, 14.2, 305.
1.3-5 Solve the following:

$$\frac{1}{2\pi} = \sqrt{\frac{386 \times (30 \times 0.00836 + 50 \times 0.0136)}{(30 \times 0.00836^2 + 50 \times 0.0136^2)}} =$$

2 · Simple Applied Problems

I. Systems of Measurement

In problem solving we have to deal with many units and their conversions to other units. The United States still uses the English system of measurement, but because all the other major industrial nations in the world are using the SI (metric) system or are converting to it, this system is coming more and more into use in this country. As a matter of fact, scientists in this country have been using the SI system for a number of years.

In this chapter we deal first with conversion of units in each of these systems and then with conversions between the two systems.

The SI system is the simpler system because everything is in multiples of 10. If one is not familiar with this system, it may seem awkward at first. However, one can readily see the advantages for calculation. Tables 2.1 and 2.2 give the more common units in the two systems. Establish

TABLE 2.1 Units of the English System

Length

1 yard (yd)		
1 foot (ft)	=	1/3 yd
1 inch (in)	=	1/12 ft
1 microinch (min)	=	0.000001 in
1 statute mile	=	1760 yd
	=	5280 ft
1 nautical mile	=	6080 ft
	=	1 minute of latitude
1 fathom	=	6 ft

Mass

1 pound (lb)		
1 ounce (oz)	=	1/16 lb
1 grain (gr)	=	1/7000 lb
1 ton	=	2000 lb

Area and Volume

1 acre	=	43,560 ft^2
1 square mile	=	640 acres
1 gallon	=	0.1337 ft^3
	=	231 in^3
1 bushel (U.S.)	=	1.2445 ft^3
	=	2150.42 in^3

orderly thought processes for converting units. For example, in converting 24 ft to inches

$$24 \text{ ft} \times \frac{12 \text{ in}}{1 \text{ ft}} = 288 \text{ in}$$

Note that the units not canceled appear in the answer.

TABLE 2.2   Units of the SI
(Metric) System

Length		
1 meter (m)		
1 decimeter (dm)	=	0.1 m
1 centimeter (cm)	=	0.01 m
1 millimeter (mm)	=	0.001 m
1 micron (μ)	=	0.000001 m
1 kilometer (km)	=	1000 m
Mass		
1 kilogram (kg)		
1 gram (g)	=	0.001 kg
1 centigram (cg)	=	0.01 g
1 milligram (mg)	=	0.001 g
Area and Volume		
10000 m^2	=	1 hectare
100 hectares	=	1 km^2
1 liter	=	1000 cm^3

The following example problems give a taste of everyday industrial problems. First we will present problems that involve conversions of units within the English system of measurement:

EXAMPLE 1.   A processing engineer needs to find the number of 6.75-in pieces that can be cut from a 10-ft, 6-in steel bar (see Fig. 2.1.). This

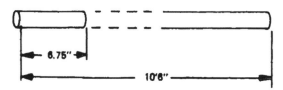

FIG. 2.1

will determine the number of steel bars needed to fill a manufacturing order. It will also affect the cost of the finished part. In solving this problem, first consider an approximation:

6.75 in $\cong$ 7 in

10 ft, 6 in $\cong$ 120 in

No. of pieces $\cong \dfrac{120 \text{ in}}{7 \text{ in}} = 17$ pieces

Given: A 10-ft, 6-in steel bar.

Required: How many 6.75-in pieces can be cut?

$$\frac{(10 \text{ ft} \times 12 \text{ in/ft}) + 6 \text{ in}}{6.75 \text{ in/piece}} = \frac{18.67 \text{ pieces}}{18 \text{ pieces} + 4.5 \text{ in}}$$

EXAMPLE 2.   A swimming pool has a volume of 10,000 ft³. What is the equivalent volume in gallons?

Required: Find the number of gallons of water needed to fill a swimming pool whose volume is 10,000 ft³.

$$\frac{10,000 \text{ ft}^3 \times 1728 \text{ in}^3/\text{ft}^3}{231 \text{ in}^3/\text{gal}} = 74,805 \text{ gal}$$

EXAMPLE 3.   In designing an air conditioning system, it is necessary to determine the capacity for removing moisture from the air. If the system removes 15 grains of water per pound of air from 6000 ft³ of air per hour, how many pounds of water are removed per hour if the density of air is 0.081 lb/ft³?

Given: An air conditioner removes 15 grains of water per pound from 6000 ft³ of air per hour. The density of air is 0.081 lb/ft³.

Required: Find the number of pounds of water removed.

$$\frac{6000 \text{ ft}^3 \times 0.081 \text{ lb/ft}^3 \times 15 \text{ gr/lb}}{7000 \text{ gr/lb}} = 1.041 \text{ lb}$$

Next let us look at similar problems in the SI system:

EXAMPLE 4.   How many 15-cm pieces can be cut from a 4-m steel bar? (See Fig. 2.2.)

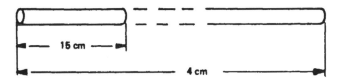

FIG. 2.2

Given: A steel bar 4 m long.

Required: Find how many 15-cm pieces can be cut.

$$\frac{4 \text{ m} \times 100 \text{ cm/m}}{15 \text{ cm}} = 26.7 \text{ pieces}$$

$$= 26 \text{ pieces } + 10.5 \text{ cm}$$

EXAMPLE 5.   How many liters are in a volume of 300 m^3?

$$\frac{300 \text{ m}^3 \times 1\text{E}6 \text{ cm}^3/\text{m}^3}{1000 \text{ cm}^3/\text{liter}} = 3\text{E}5 \text{ liters}$$

II. English-SI Conversions

Many of the larger American corporations because of the international nature of their business have converted to the SI system of measurement. Because both systems are used in business, it is frequently necessary to make conversions from one system to the other. Refer to Table 2.3 for conversions between the two systems.

Two of the calculators being discussed have routines for converting from English to metric and metric to English. The TI-68 converts from English to metric as follows:

To convert 1 in to cm, press 2nd *in-cm* =. To convert 1 cm to in, press *INV 2nd in-cm* =.

For the HP-32SII the procedure for converting 1 in to cm is as follows: Press ⌐ → *cm*. To convert 1 cm to in, press Γ → *in*.

Following are some English-SI conversion problems:

EXAMPLE 6.   We are accustomed to seeing automobile engine displacement ratings expressed in cubic inches. Now we are frequently seeing them expressed in liters. To make comparisons it is necessary to convert from one system to the other. An engine is rated at a displacement of 2 liters. Convert to cubic inches.

TABLE 2.3 Equivalent English and Metric Units

English		Metric	Metric		English
1 lb	=	453.5924 g	1 kg	=	2.205 lb
1 in	=	2.54 cm	1 cm	=	0.3937 in
1 ft	=	30.48 cm	1 m	=	39.37 in
				=	1.09 yd
1 mi	=	1.609 km	1 km	=	0.6214 mi
				=	3281 ft
1 qt	=	0.94607 liter	1 liter	=	1.057 qt
1 gal	=	3.785 liter		=	0.2642 gal
				=	61.024 in³
1 acre	=	0.40469 hectare	1 hectare	=	2.471 acres
1 mi²	=	2.590 km²	1 km²	=	0.3861 mi²

Given: An engine is rated at 2-liter displacement.

Required: Convert displacement to cubic inches.

$$2 \text{ liter} \times 61.024 \text{ in}^3/\text{liter} = 122.05 \text{ in}^3$$

EXAMPLE 7.   In a similar manner, we need to make conversions to compare performance. A car achieves an average performance of 18 mi/gal. Convert to kilometers per liter.

Given: A car goes 18 mi/gal.

Required: Convert to km/liter

$$\frac{18 \text{ mi/gal} \times 1.609 \text{ km/mi}}{3.785 \text{ liter/gal}} = 7.652 \text{ km/liter}$$

EXAMPLE 8.   Here is a problem involving area comparisons between the two systems. A farmer plants 3 bushels per acre. Convert to cubic meters per hectare.

Given: A farmer plants 3 bushels of corn per acre.

Required: Convert to cubic meters per hectare.

$$\frac{3 \text{ bu/acre} \times 0.03524 \text{ m}^3/\text{bu}}{0.40469 \text{ ha/acre}} = 0.2612 \text{ m}^3/\text{ha}$$

III. Plane Geometric Straight-Sided Areas

Technical work involves a great deal of calculation of areas. You should be familiar with the formulas for most of these from the study of high school geometry. Formulas for calculating the areas of the most common straight-sided figures appear in Fig. 2.3. It should be noted that the area

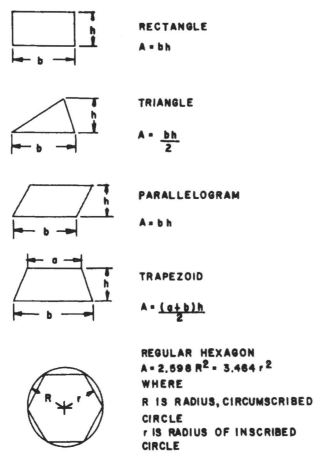

RECTANGLE

$A = bh$

TRIANGLE

$A = \dfrac{bh}{2}$

PARALLELOGRAM

$A = bh$

TRAPEZOID

$A = \dfrac{(a+b)h}{2}$

REGULAR HEXAGON
$A = 2.598 R^2 = 3.464 r^2$
WHERE
R IS RADIUS, CIRCUMSCRIBED CIRCLE
r IS RADIUS OF INSCRIBED CIRCLE

FIG. 2.3   Areas of plane straight-sided figures.

of a triangle is equal to half the product of the base and the height regardless of whether it is a right or oblique triangle. Also note that the area of a regular hexagon is composed of six equilateral triangles whose height is r and whose base is 1.1547r.

IV. Areas and Circumferences of Circles

Figure 2.4 gives the formulas for calculating the area of a number of circular figures (Oberg and Jones, 1964). You should remember from high school geometry that the area of a circle is equal to πr^2. In most practical applications you will be dealing with the diameter of a circle rather than the radius. Therefore it is more convenient to use the formula

$$A = \frac{\pi d^2}{4} \tag{1}$$

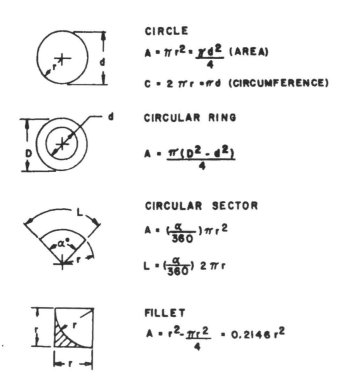

FIG. 2.4 Circular areas.

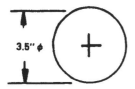

FIG. 2.5

The value of π, the circular constant, is a number that can be carried out to any number of decimal places without the number sequence repeating itself. The π key on your calculator will enter the value of π to eight decimal places. For the TI-68 calculator it is necessary to press the 2nd key before pressing the π key. For the HP-32SII press $\lceil \pi$.

The determination of circular areas is one of the most common calculations involved in the solution of engineering problems. It is used in the measurement of strength of materials and in the calculation of weight of structural members. Determination of circumferences of circles is used in all sorts of calculations involving the speed of machine elements.

Following are some examples of circular problems:

EXAMPLE 9. Calculate the circumference of a 3.5-in diameter circle and its area.

Required: Calculate area and circumference of circle shown in Fig. 2.5

$$A = \frac{\pi d^2}{4} = \frac{\pi \times 3.5^2 \text{ in}}{4} = 9.621 \text{ in}^2$$

$$C = \pi d = \pi \times 3.5 \text{ in} = 10.99 \text{ in}$$

EXAMPLE 10. Calculate the cross-sectional area of a tube with an outside diameter of 2 in and an inside diameter of 1.5 in.

Required: Calculate cross-sectional area of tube shown in Fig. 2.6.

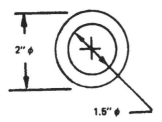

FIG. 2.6

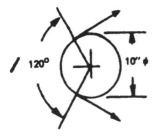

FIG. 2.7

$$A = \frac{\pi(D^2 - d^2)}{4} = \frac{\pi(2^2 - 1.5^2)}{4} = 1.374 \text{ in}^2$$

EXAMPLE 11. Calculation of the length of a circular sector is involved in determining the length of belts and chains. A belt wraps around 120° on a 10-in diameter pulley. Calculate the length of wrap. (See Fig. 2.7.)

Given: A belt wraps 120° on a 10-in diameter pulley.

Required: Calculate length of wrap.

$$L = \frac{\alpha \pi d}{360°} = \frac{120° \times \pi \times 10 \text{ in}}{360°} = 10.47 \text{ in}$$

EXAMPLE 12. Most castings and forgings as well as standard rolled structural members have fillets or radii at the intersections of planes or other surfaces. Often the weight of material in fillets is significant. Calculate the area of a fillet whose radius is 0.5 in. (See Fig. 2.8.)

Required: Calculate the area under a 0.5-in radius fillet.

$$A = 0.2146r^2 = 0.2146 \times 0.5^2 \text{ in} = 0.0536 \text{ in}^2$$

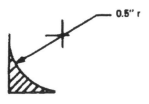

FIG. 2.8

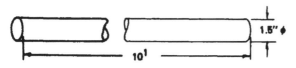

FIG. 2.9

Remember that the volume of any figure with a uniform cross section is equal to the product of its length and cross-sectional area.

Volume of a cylinder: $V = \dfrac{\pi d^2 L}{4}$ (2)

Volume of a tube: $V = \dfrac{\pi (D^2 - d^2)L}{4}$ (3)

These formulas are used in determining the weight of round bars and tubes, as in the following examples.

EXAMPLE 13. Calculate the weight of a 1.5-in diameter steel bar 10 ft long. The density of steel is 0.283 lb/in³ (D − wt/vol). (See Fig. 2.9.)

Given: A 1.5-in ϕ steel bar is 10 ft long. The density of steel is 0.283 lb/in³.

Required: Calculate weight of bar.

$$V = \frac{\pi d^2 L}{4} = \frac{\pi \times 1.5^2 \text{ in} \times 10 \text{ ft} \times 12 \text{ in/ft}}{4} = 212.06 \text{ in}^3$$

$W = VD = 212.06 \text{ in}^3 \times 0.283 \text{ lb/in}^3 = 60.01 \text{ lb}$

EXAMPLE 14. To calculate the speed of a rolling wheel, it is necessary to consider the circumference (C). The diameter of the automobile wheel shown in Fig. 2.10 is 30 in. Calculate the revolutions per minute when the automobile is traveling at the legal speed limit of 55 mi/hr.

Given: A 30-in diameter wheel is rolling at 55 mph.

Required: Calculate rpm.

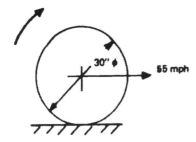

FIG. 2.10

$$C = \pi d = \frac{\pi \times 30 \text{ in}}{12 \text{ in/ft}} = 7.854 \text{ ft/rev}$$

$$D = \frac{55 \text{ mi/hr} \times 5280 \text{ ft/mi}}{60 \text{ min/hr}} = 4840 \text{ ft/min}$$

$$\frac{4840 \text{ ft/min}}{7.854 \text{ ft/rev}} = 616.2 \text{ rpm}$$

V. Electrical Problems

The three basic quantities used in electrical or electronic calculations are current, voltage, and resistance. The unit for current is the ampere (abbreviated A or amp), which corresponds to a flow of 6.24E18 electrons per second. The symbol for current is I (Harris and Hemmerling, 1972).

The unit of electrical pressure is the volt (abbreviated V). It is also referred to as *potential* or *electromotive force*. The symbol for potential is E.

The unit of electrical resistance is the ohm (abbreviated Ω). It is defined as the resistance of a column of pure mercury 1 mm^2 in cross section and 106.3 cm long at a temperature of 0° C. The symbol for resistance is R.

The relationship between these three quantities is expressed by Ohm's law:

$$I = \frac{E}{R} \tag{4}$$

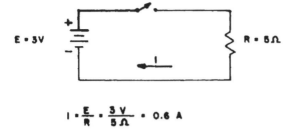

$$I = \frac{E}{R} = \frac{3 \text{ V}}{5 \Omega} = 0.6 \text{ A}$$

FIG. 2.11 Simple dc circuit.

The unit for electrical power is the watt (abbreviated W). The symbol for power is P. The relationship of power to the other electrical units may be expressed by any of the following equations:

$$P = EI \tag{5}$$

$$P = I^2R \tag{6}$$

$$P = \frac{E^2}{R} \tag{7}$$

In this chapter we deal only with direct current circuits involving resistances in series. Circuits involving resistances in parallel are taken up in Section II of the next chapter. A simple dc circuit is shown in Fig. 2.11.

In electronics applications it is important not to exceed the wattage ratings of resistors. The following equations demonstrate the calculation of safe voltage and current for the 30-kΩ resistor, rated at 1 W, shown in Fig. 2.12 (Malmstadt et al., 1963).

Given: The Circuit shown in Fig. 2.12.

FIG. 2.12 Current and voltage for 1-W rating resistor.

FIG. 2.13 dc resistances in series.

Required: (a) Calculate the safe current.

$$I = \sqrt{\frac{P}{R}} = \sqrt{\frac{1\ W}{30,000\ \Omega}} = 0.00577\ A \text{ or } 5.77\ mA$$

(b) Calculate safe voltage.

$$E = \sqrt{PR} = \sqrt{1\ W \times 30,000\ \Omega} = 173.2\ V$$

When resistances are in series, the resistance of the complete circuit is equal to the sum of the individual resistances. The sum of the IR drops across the individual resistances must be equal to the source voltage.

Given: Circuit data as shown in Fig. 2.13.

Required: Calculate R, I, IR_2, IR_1.

$$R = R_1 + R_2 = 10\ \Omega + 20\ \Omega = 30\ \Omega$$

$$I = \frac{E}{R} = \frac{6\ V}{30\ \Omega} = 0.2\ A$$

$$IR_1 = 0.2\ A \times 10\ \Omega = 2\ V$$

$$IR_2 = 0.2\ A \times 20\ \Omega = 4\ V$$

VI. Chemistry Problems

The fundamental unit used in chemical calculations is the atomic weight of an element. It is a number which compares the mass of an average atom of that element to that of carbon-12, which is assigned a mass of exactly 12. Thus the atomic weights of the various elements are the relative masses of the average atoms of those elements. *The mass in grams of an element equal to its atomic weight is called one gram atomic weight of the element* (Sorum, 1969).

It has been established that a gram atomic weight of any element contains 6.023E23 atoms (Avogadro's number). This quantity is referred to as one mole. The mole is a term that is also applied to molecules and groups of atoms. One mole of a substance is 6.023E23 units of that substance. For atomic weights of the various chemical elements, refer to Table 2.4.

The sum of the atomic weights represented by the formula of a substance is its formula weight. The formula weight of sodium chloride (NaCl) is equal to the sum of the atomic weights of Na and Cl:

22.99 g Na + 35.45 g Cl = 58.44 g NaCl

The formula weight of sulfuric acid (H_2SO_4) is

2×1.008 g H $+ 32.06$ g S $+ 4 \times 16$ g O $= 98.08$ g H_2SO_4

The formula weights in the preceding examples are the weights in grams per mole of the substances.

Some example problems follow:

EXAMPLE 15. A compound contains 27.27% carbon and 72.73% oxygen. Determine its formula.

Solution:

$$\frac{27.27 \text{ g C}}{12 \text{ g/mole C}} = 2.2725 \text{ moles C}$$

$$\frac{72.73 \text{ g O}}{16 \text{ g/m O}} = 4.54 \text{ moles O}$$

$$\frac{4.54}{2.27} = 2.0 \text{ (ratio of moles O to moles C)}$$

Therefore the formula is CO_2.

EXAMPLE 16. A chemical compound when analyzed contained 621.6 g Pb and 64.0 g O. Calculate the empirical formula.

TABLE 2.4 Atomic Weights of Elements Referred to $^{12}C = 12.0000$

Element	Symbol	Atomic Weight	Element	Symbol	Atomic Weight
Actinium	Ac	(227)	Mercury	Hg	200.59
Aluminum	Al	26.9815	Molybdenum	Mo	95.94
Americium	Am	(243)	Neodymium	Nd	144.24
Antimony	Sb	121.75	Neon	Ne	20.183
Argon	Ar	39.948	Neptunium	Np	(237)
Arsenic	As	74.9216	Nickel	Ni	58.71
Astatine	At	(210)	Niobium	Nb	92.906
Barium	Ba	137.34	Nitrogen	N	14.0067
Berkelium	Bk	(249)	Osmium	Os	190.2
Beryllium	Be	9.0122	Oxygen	O	15.9994
Bismuth	Bi	208.980	Palladium	Pd	106.4
Boron	B	10.811	Phosphorus	P	30.9738
Bromine	Br	79.909	Platinum	Pt	195.09
Cadmium	Cd	112.40	Plutonium	Pu	(242)
Calcium	Ca	40.08	Polonium	Po	(210)
Californium	Cf	(251)	Potassium	K	39.102
Carbon	C	12.01115	Praseodymium	Pr	140.907
Cerium	Ce	140.12	Promethium	Pm	(145)
Cesium	Cs	132.905	Protactinium	Pa	(231)
Chlorine	Cl	35.453	Radium	Ra	(226)
Chromium	Cr	51.996	Radon	Rn	(222)
Cobalt	Co	58.9332	Rhenium	Re	186.2
Copper	Cu	63.54	Rhodium	Rh	102.905
Curium	Cm	(247)	Rubidium	Rb	85.47
Dysprosium	Dy	162.50	Ruthenium	Ru	101.07
Einsteinium	Es	(254)	Samarium	Sm	150.35

TABLE 2.4 Continued

Element	Symbol	Atomic Weight	Element	Symbol	Atomic Weight
Erbium	Er	167.26	Scandium	Sc	44.956
Europium	Eu	151.96	Selenium	Se	78.96
Fermium	Fm	(253)	Silicon	Si	28.086
Fluorine	F	18.9984	Silver	Ag	107.870
Francium	Fr	(223)	Sodium	Na	22.9898
Gadolinium	Gd	157.25	Strontium	Sr	87.62
Gallium	Ga	69.72	Sulfur	S	32.064
Germanium	Ge	72.59	Tantalum	Ta	180.948
Gold	Au	196.967	Technetium	Tc	(99)
Hafnium	Hf	178.49	Tellurium	Te	127.60
Helium	He	4.0026	Terbium	Tb	158.924
Holmium	Ho	164.930	Thallium	Tl	204.37
Hydrogen	H	1.00797	Thorium	Th	232.038
Indium	In	114.82	Thulium	Tm	168.934
Iodine	I	126.9044	Tin	Sn	118.69
Iridium	Ir	192.2	Titanium	Ti	47.90
Iron	Fe	55.847	Tungsten	W	183.85
Krypton	Kr	83.80	Uranium	U	238.03
Lanthanum	La	138.91	Vanadium	V	50.942
Lead	Pb	207.19	Xenon	Xe	131.30
Lithium	Li	6.939	Ytterbium	Yb	173.04
Lutetium	Lu	174.97	Yttrium	Y	88.905
Magnesium	Mg	24.312	Zinc	Zn	65.37
Manganese	Mn	54.9380	Zirconium	Zr	91.22
Mendelevium	Md	(256)			

Solution:

$$\frac{621.6 \text{ g Pb}}{207.19 \text{ g/mole}} = 3 \text{ moles Pb}$$

$$\frac{64 \text{ g O}}{16 \text{ g/mole}} = 4 \text{ moles O}$$

The formula is Pb_3O_4.

EXAMPLE 17. Calculate the formula weight of H_3PO_4.

Solution:

$$\begin{array}{ccc} \text{(g/mole H)} & \text{(g/mole P)} & \text{(g/mole O)} \\ 3 \times 1 & + \quad 30.97 & + \quad 4 \times 16 \quad = 97.97 \text{ g/mole } H_3PO_4 \end{array}$$

EXAMPLE 18. Calculate the mass of 0.35 mole of $CaCO_3$.

$$\begin{array}{ccc} \text{(g/mole Ca)} & \text{(g/mole C)} & \text{(g/mole O)} \\ 0.35 \text{ m}(40.08 + & 12.01 & + \quad (3 \times 16 \quad = 35.03 \text{ g} \end{array}$$

EXAMPLE 19. How many grams of copper are in 500 g of $CuSO_4$?

Solution:

$$\begin{array}{ccc} \text{(g/mole Cu)} & \text{(g/mole S)} & \text{(g/mole O)} \\ 63.54 & + \quad 32.06 & + \quad 4 \times 16 \end{array}$$

$$= 159.6 \text{ g/mole formula weight } CuSO_4$$

$$\frac{500 \text{ g}}{159.6 \text{ g/mole } CuSO_4} \times 63.54 \text{ g/mole Cu} = 199.1 \text{ g Cu}$$

EXAMPLE 20. Calculate the percentage of oxygen in $KClO_3$.

Solution:

$$\frac{3 \times 16 \text{ g O} \times 100}{(30.1 + 35.45 + 3 \times 16) \text{ g } KClO_3} = 42.27\%$$

EXAMPLE 21. 500 g of NaOH is reacted upon by H_2SO_4 according to the following equation:

$$2NaOH + H_2SO_4 \rightarrow Na_2SO_4 + 2H_2O$$

How much H_2SO_4 is required?

Solution: Calculating formula weights:

$22.99 + 16 + 1 = 39.99$ g/mole NaOH

$2 \times 1 + 32.06 + 4 \times 16 = 98.06$ g/mole H_2SO_4

By proportion:

$$\frac{2 \times 39.99}{500 \text{ g}} \text{ g/mole NaOH} = \frac{98.06}{x} \text{ g/mole } H_2SO_4$$

$$x = 613 \text{ gm } H_2SO_4$$

Other chemistry problems will be taken up in later chapters as further problem-solving techniques are developed.

Exercise 2.1

2.1-1   Water weighs 62.4 lb/ft³. How much does a gallon of water weigh?

2.1-2   A water tank contains 500,000 gal of water. How much does the water weigh?

2.1-3   How many 4.75-in pieces can be cut from a 12-ft, 6-in bar?

2.1-4   An acre of land contains 43,560 ft². If a square plot of land has an area of 1 acre, what are its dimensions in feet?

2.1-5   A grain bin has a volume of 1000 ft³. How many bushels does it hold?

2.1-6   A square plot of land has an area of 5 ha. What are its dimensions in meters?

2.1-7   How many 25.5-cm pieces can be cut from a 3.5-m steel bar?

2.1-8   How far apart are two points if it takes 2 hr, 10 min to drive between them at 90 km/hr?

2.1-9   How many liters are there in a volume of 125 m³?

2.1-10 A liter of water weighs 1 kg. What is the weight of water in a tank that contains 150 m³?

Exercise 2.2

2.2-1   Convert 550 mi/hr to meters per second.

2.2-2   The present legal speed limit in many states is 55 mi/hr. What is the equivalent in kilometers per hour?

2.2-3   A performance of 12 km/liter is claimed for a foreign car. What is the equivalent in miles per gallon?

2.2-4   A square plot of land has an area of 10 ha. Convert to acres and solve for the dimensions of the plot in feet.

2.2-5   A field of Iowa corn produces 125 bushels per acre. What is the equivalent yield in cubic meters per hectare?

2.2-6   Drugs are weighed in grains in the English system. The common aspirin tablet weighs 5 grains. What is the equivalent weight in grams?

2.2-7   An automobile engine has a displacement of 325 in³. What is the equivalent displacement in liters?

2.2-8   If you are at 42°N latitude, what is your distance from the equator in kilometers? (Hint: One degree of latitude is a nautical mile.)

2.2-9   The density of steel is 0.283 lb/in³. Convert to kg/cm³.

2.2-10 Water flows through a channel at the rate of 3000 ft³/min. Convert to liters per second.

Exercise 2.3

2.3-1   Calculate the rpm of an 84-in diameter tractor wheel for 5 mi/hr.

2.3-2   Calculate the weight of a 1.25-in-diameter steel bar that is 16 ft long. The density of steel is 0.283 lb/in³.

2.3-3   Calculate the weight of a 1.5-in (across flats) hexagonal steel bar that is 12 ft long.

2.3-4   Calculate the weight of a 3-in O.D. × 2-in I.D. steel tube that is 12 ft long.

2.3-5   Calculate the number of acres in the plot of ground shown in Fig. 2.14.

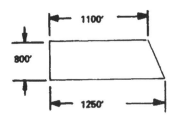

FIG. 2.14

2.3-6   Calculate the length of the 140° arc in Fig. 2.15.

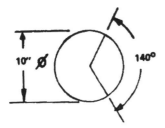

FIG. 2.15

2.3-7   Calculate the cross-sectional area of the beam shown in Fig. 2.16.

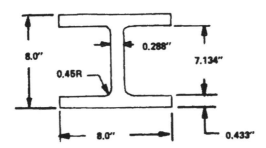

FIG. 2.16

2.3-8   Calculate the cross-sectional area of the angle shown in Fig. 2.17.

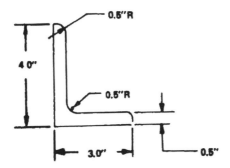

FIG. 2.17

2.3-9   The steel plate shown in Fig. 2.18 is 0.5 in thick. Calculate its area and weight.

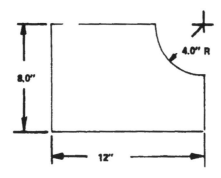

FIG. 2.18

2.3-10 A cylindrical water tank 10 ft in diameter and 40 ft high is filled with water weighing 62.4 lb/ft³. Determine the weight of water pressing on the bottom.

Exercise 2.4

2.4-1 A 10-kΩ resistor has a rating of 0.5W. (a) Calculate the safe current. (b) Calculate the safe voltage.

2.4-2 A 40-kΩ resistor has a rating of 1.5W. (a) Calculate the safe current. (b) Calculate the safe voltage.

2.4-3 A current of 5mA flows through a 35-kΩ resistor. Calculate the power in watts.

2.4-4 Current at a potential of 115 V flows through a 48-kΩ resistor. Calculate the power in watts.

2.4-5 If the power in a circuit is 1 W and the potential is 115 V, calculate the current.

2.4-6 A circuit has two resistances connected in series. R_1 is 5 Ω and R_2 is 10 Ω. The potential is 12 V. Calculate the IR drop across each resistance.

2.4-7 A current of 0.5 A at a potential of 10 V flows through a circuit containing two resistances connected in series. The IR drop through R_1 is 3 V. Calculate IR_2, R_1, and R_2.

2.4-8 A current at a potential of 12 V flows through a circuit containing two resistances connected in series. If R_1 is 5 Ω and IR_1 is 4 V, calculate IR_2, R_2, and I.

2.4-9 Ten dry cells, each with an emf of 1.5 V and an internal resistance of 0.2 Ω are connected in series. How much current will flow when an external load of 0.5 Ω is applied?

2.4-10   A 12-V storage battery with 1.2 Ω of internal resistance is to
be charged from a 24-V circuit. What resistance must be placed
in series with the battery if the charging rate is 5 A?

Exercise 2.5

2.5-1   A compound contains 52.3% carbon, 13% hydrogen, and
34.7% oxygen. Calculate the empirical formula of the compound.

2.5-2   Calculate the formula weight of Na_2SO_4.

2.5-3   Calculate the mass of 6 moles of H_2SO_4.

2.5-4   How many grams of oxygen are in 550 g of $KClO_3$?

2.5-5   Calculate the percentage of chromium in $K_2Cr_2O_7$.

2.5-6   500 g of $Al(OH)_3$ is reacted upon by H_2SO_4 according to the
following equation:

$$2Al(OH)_3 + 3H_2SO_4 \rightarrow Al_2(SO_4)_3 + 6H_2O.$$

What weight of H_2SO_4 is required?

3 · Reciprocals, Powers, and Roots

I. Reciprocals

The reciprocal of a number is equal to 1 divided by the number. For algebraic notation the "x^{-1}" key enables one to calculate the reciprocal of a number. For the TI-68 it is necessary to press the "=" key after "x^{-1}." For the FX-7000GA the "EXE" key must be pressed after the "x^{-1}." If the number is part of an equation, it is not necessary to press the "=" key until the equation has been completed.

For the HP-32SII the "$\frac{1}{x}$" key is used to calculate the reciprocal.

For example, $\frac{1}{3.2} = 0.3125$.

Calculator routine for algebraic notation:

PRESS	DISPLAY
3.2 $\underline{x^{-1}}$ =	0.3125

Calculator routine for reverse Polish notation:

PRESS	DISPLAY
$3.2 \underset{x}{\underline{\frac{1}{}}}$	0.3125

One convenience of the reciprocal function is the ability to use a number in the display which is to be applied in the denominator in a following calculation. This can be accomplished by pressing the 1/x key and then using the result as a multiplier.

EXAMPLE 1. The number 5.125 is in the display as the result of a previous calculation. The next calculation is

$$\frac{7.36 \times 8.45}{5.125}$$

Calculator routine for algebraic notation:

PRESS	DISPLAY
	5.125
$x^{-1} \times 7.36$	
$\times\ 8.45\ =$	12.135

Calculator routine for reverse Polish notation:

PRESS	DISPLAY
	5.125
$\frac{1}{x}\ 7.36 \times$	
$8.45 \times$	12.135

If we are making conversions from SI to English units and we know the English to SI conversion, we can use the reciprocal of the known conversion.

EXAMPLE 2. It is known that 1 in = 2.54 cm. What is the conversion from centimeters to inches?

Calculator routine:

ENTER	PRESS	DISPLAY
2.54	1/x	0.3937008

II. Problems Involving Reciprocals

Many scientific problems involve solving for a sum of reciprocals.

EXAMPLE 3.   Solve for x: $\dfrac{1}{3.4} + \dfrac{1}{7.25} = \dfrac{1}{x}$.

Calculator routine for algebraic notation:

PRESS	DISPLAY
3.4 x^{-1} +	.
7.25 x^{-1} =	0.432
x^{-1} =	2.315

Calculator routine for reverse Polish notation:

PRESS	DISPLAY
3.4 $\dfrac{1}{x}$	
7.25 $\dfrac{1}{x}$ +	
$\dfrac{1}{x}$	2.315

A typical application of this type of equation is the calculation of the net resistance of a group of parallel electrical circuits where the resistance of each individual circuit is known. The sum of the reciprocals of the individual resistances is equal to the reciprocal of the resistance of the entire circuit.

$$\frac{1}{R_T} = \frac{1}{R_1} + \frac{1}{R_2} + \frac{1}{R_3} \tag{1}$$

where

R_T = total resistance of the parallel circuit

R_1, R_2, R_3 = parallel resistances

EXAMPLE 4. Three circuits having resistances of 25, 75, and 125 Ω, respectively, are connected in parallel. Calculate the total resistance of the combined circuit. (See Fig. 3.1.)

Given: Three resistances are connected in parallel as shown.

Required: Calculate the total resistance of the circuit.

$$\frac{1}{R_T} = \frac{1}{25\ \Omega} + \frac{1}{75\ \Omega} + \frac{1}{125\ \Omega}; R_T = 16.3\ \Omega$$

Calculator routine for algebraic notation:

PRESS	DISPLAY
25 $\underline{x^{-1}}$ +	
75 $\underline{x^{-1}}$ +	
125 $\underline{x^{-1}}$ =	0.061
$\underline{x^{-1}}$ =	16.3

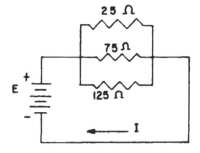

FIG. 3.1

Calculator routine for reverse Polish notation:

PRESS	DISPLAY
$25 \dfrac{1}{x}$	
$75 \dfrac{1}{x} +$	0.061
$\dfrac{1}{x}$	16.304

For a series-parallel circuit, the procedure is to calculate the total resistance of each group of parallel resistances and then to add these total resistances where the groups are connected in series. For solving this type of problem, good use can be made of the memories in your calculator (Malmstadt et al., 1963).

EXAMPLE 5. Calculate the total resistance of the network shown in Fig. 3.2, and calculate the current flowing through the different branches of the network. Calculate the IR drops across A, B, and C.

Given: Network shown in Fig. 3.2.

Required: (a) net resistance, (b) current through each branch, (c) IR drops across A, B, and C.

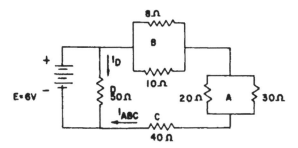

FIG. 3.2

$$\frac{1}{R_A} = \frac{1}{20\ \Omega} + \frac{1}{30\ \Omega} \qquad R_A = 12\ \Omega$$

$$\frac{1}{R_B} = \frac{1}{8\ \Omega} + \frac{1}{10\ \Omega} \qquad R_B = 4.44\ \Omega$$

$$R_{ABC} = R_A + R_B + B_C = 12\ \Omega + 4.44\ \Omega + 40\ \Omega = 56.44\ \Omega$$

$$\frac{1}{R} = \frac{1}{56.44\ \Omega} + \frac{1}{50\ \Omega} \qquad R = 26.51\ \Omega$$

$$I_D = \frac{E}{R_D} = \frac{6\ V}{50\ \Omega} = 0.12\ A$$

$$I_{ABC} = \frac{E}{R_{ABC}} = \frac{6\ V}{56.44\ \Omega} = 0.106\ A$$

$$I_{ABC}R_B = 0.106\ A \times 4.44\ \Omega = 0.472\ V$$

$$I_{ABC}R_A = 0.106\ A \times 12\ \Omega = 1.275\ V$$

$$I_{ABC}R_C = 0.106\ A \times 40\ \Omega = 4.252\ V$$

Calculator routine for algebraic notation:

PRESS	DISPLAY	REMARKS
20 x^{-1} +		
30 x^{-1} = x^{-1} =	12.00	RA
STO ALPHA A =		
STO ALPHA A1 =		
8 x^{-1} +		
10 x^{-1} = x^{-1} =	4.44	RB
STO ALPHA B =		
STO + ALPHA A =		
40 STO + ALPHA A =		
RCL ALPHA A =	56.44	RABC
x^{-1} +		
50 x^{-1} = x^{-1} =	26.51	R

PRESS	DISPLAY	REMARKS
6 + $\underline{ALPHA}$ A =	0.106	IABC
$\underline{STO}$ $\underline{ALPHA}$ D =		
× $\underline{ALPHA}$ B =	0.472	IRB
$\underline{ALPHA}$ D × $\underline{ALPHA}$ A1 =	1.276	IRA
40 × $\underline{ALPHA}$ D =	4.252	IRC

Calculator routine for reverse Polish notation:

PRESS	DISPLAY	REMARKS
20 $\frac{1}{\underline{x}}$	0.05	
30 $\frac{1}{\underline{x}}$ + $\frac{1}{\underline{x}}$	12.00	RA
$\underline{STO}$ A $\underline{STO}$ B 8 $\frac{1}{\underline{x}}$	0.125	
10 $\frac{1}{x}$ + $\frac{1}{x}$ ·	4.44	RB
$\underline{STO}$ C $\underline{STO+}$ A		
40 $\underline{STO+}$ A $\underline{RCL}$ A	56.44	RABC
$\frac{1}{\underline{x}}$ 50 $\frac{1}{\underline{x}}$ + $\frac{1}{\underline{x}}$ ·	26.51	R
6 $\underline{RCL}$ A ÷ $\underline{STO}$ D	0.106	IABC
$\underline{RCL}$ C $\underline{RCL}$ D ×	0.472	IRB
$\underline{RCL}$ D $\underline{RCL}$ B ×	1.276	IRA
40 $\underline{RCL}$ D ×	4.252	IRC

This type of equation is also widely used in the solution of optical problems. The basic equation is

$$\frac{1}{D_o} + \frac{1}{D_i} = \frac{1}{f} \tag{2}$$

where

D_o = object distance from the lens
D_i = image distance from the lens
f = focal length of the lens

EXAMPLE 6.   If the camera lens shown in Fig. 3.3 has a focal length of 4 cm and the object distance is 500 cm, solve for the image distance from the lens. The equation may be transposed as follows:

$$\frac{1}{D_i} = \frac{1}{f} - \frac{1}{D_o}$$

Given: Focal length of camera lens is 4 cm.

Required: Determine image distance if object distance is 500 cm.

$$\frac{1}{D_o} + \frac{1}{D_i} = \frac{1}{f}$$

$$\frac{1}{D_i} = \frac{1}{f} - \frac{1}{D_o} = \frac{1}{4 \text{ cm}} - \frac{1}{500 \text{ cm}}$$

$$D_i = 4.032 \text{ cm}$$

III. Powers and Roots

The quick determination of powers and roots is essential in solving many scientific problems. The operation of the $\sqrt{x}$ and the x^2 keys has already

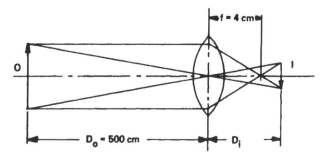

FIG. 3.3

been explained (see Chapter 1). Your calculator also has the capability of determining any power or root of a number. To find the x power of a number, use the y^x key. For a calculator with algebraic logic, the procedure is as follows:

EXAMPLE 7. $2^5 = 32$.

Calculator routine for algebraic notation:

ENTER	PRESS	DISPLAY
2	y^x	
5	=	32

Calculator routine for reverse Polish notation:

PRESS	DISPLAY
2 ENTER	
5 y^x	32

The TI-68 uses the $\sqrt[x]{y}$ key to solve for any root of a number.

EXAMPLE 8. $\sqrt[3]{125} = 5$.

Calculator routine for algebraic notation:

PRESS	DISPLAY
3 2nd $\sqrt[x]{y}$	
125 =	5

The HP-32SII uses the $\sqrt[x]{y}$ key. The following is the solution of the preceding problem with this calculator:

Calculator routine for reverse Polish notation:

PRESS	DISPLAY
125 ENTER	
3 $\sqrt[x]{y}$	5

Sums of powers and roots may be entered in much the same manner as sums of products.

EXAMPLE 9. $3^2 + 2^5 - 7^2 = -8$.

Calculator routine for algebraic notation:

ENTER	PRESS	DISPLAY
3	x^2 +	
2	y^x	
5	−	
7	x^2 =	−8

Calculator routine for reverse Polish notation:

PRESS	DISPLAY
3 ⌐ x^2	9
2 <u>ENTER</u> 5 y^x +	41
7 ⌐ $\underline{x^2}$ −	−8

IV. Problems Involving Powers and Roots

We will present a number of applications of powers and roots to give some idea of the mathematical manipulations involved. These applications are explained more fully in more advanced courses.

In the study of algebra there are a number of problems involving the solution of quadratic equations. The typical quadratic equation has the form (Washington, 1970):

$$ax^2 + bx + c = 0$$

The value of x can be determined by application of the quadratic formula:

$$x = \frac{-b \pm \sqrt{b^2 - 4ac}}{2a} \tag{3}$$

Note that this equation can yield two roots.

EXAMPLE 10. For the equation $x^2 - 20x - 100 = 0$, solve for the value of x.

$$x = \frac{-(-20) \pm \sqrt{20^2 - (4)(1)(-100)}}{(2)(1)} = 10 \pm 10\sqrt{2}$$

Calculator routine for algebraic notation:

PRESS	DISPLAY
1 <u>STO</u> <u>ALPHA</u> A =	
<u>(−)</u> 20 <u>STO</u> <u>ALPHA</u> B =	
<u>(−)</u> 100 <u>STO</u> <u>ALPHA</u> C =	
$\sqrt{\ }$ (<u>ALPHA</u> B x^2 −4 ×	
<u>ALPHA</u> C) = <u>STO</u> <u>ALPHA</u> D =	
(<u>(−)</u> <u>ALPHA</u> B + <u>ALPHA</u> D)	
÷ 2 × <u>ALPHA</u> A =	24.1421
(<u>(−)</u> <u>ALPHA B</u> − <u>ALPHA</u> D)	
÷ 2 × <u>ALPHA</u> A =	−4.1421

Calculator routine for reverse Polish notation:

PRESS	DISPLAY
1 <u>STO</u> A 20 <u>+/−</u> <u>STO</u> B	
100 <u>+/−</u> <u>STO</u> C	
<u>RCL</u> B ↰ x^2 <u>ENTER</u> 4	
<u>RCL</u> A × <u>RCL</u> C × −	
$\sqrt{x}$ <u>STO</u> D	
<u>RCL</u> B <u>+/−</u> <u>RCL</u> D + 2	
÷ RCL A ÷	24.1421
<u>RCL</u> B <u>+/−</u> <u>RCL</u> D − 2	
÷ RCL A ÷	−4.1421

In the study of strength of materials frequent use is made of a property of beam sections which is called the *moment of inertia*. This is designated by the symbol I, and its units are in^4. For a rectangular section (Bassin et al., 1969):

$$I = \frac{bh^3}{12} \tag{4}$$

where

    b = width of the section, in
    h = height of the section, in

For a circular cross section:

$$I = \frac{\pi d^4}{64} \tag{5}$$

where

    d = diameter, in

For a tubular cross section:

$$I = \frac{\pi(d_o^4 - d_i^4)}{64} \tag{6}$$

where

    d_o = outside diameter, in
    d_i = inside diameter, in

EXAMPLE 11. Calculate the moment of inertia for a rectangular beam with b = 4 in and h = 8 in.

$$I = \frac{bh^3}{12} = \frac{4 \text{ in} \times (8 \text{ in})^3}{12} = 170.67 \text{ in}^4$$

Calculator routine for algebraic notation:

ENTER	PRESS	DISPLAY
4	× (	
8	y^x	
3	) ÷	512
12	=	170.67

Parentheses are necessary if the equation is entered in the order shown.

Calculator routine for reverse Polish notation:

PRESS	DISPLAY
8 ENTER 3 y^x	512
4 ×	2048
12 ÷	170.67

EXAMPLE 12. Calculate the moment of inertia of a tubular section having an O.D. of 3 in and an I.D. of 2.25 in.

$$I = \frac{\pi(d_o^4 - d_i^4)}{64} = \frac{\pi(3^4 - 2.25^4)}{64} = 2.718 \text{ in}^4$$

Calculator routine for algebraic notation:

ENTER	PRESS	DISPLAY
2nd π	× (	
3	y^x	
4	−	
2.25	y^x	
4	) ÷	
64	=	2.718

Calculator routine for reverse Polish notation:

PRESS	DISPLAY
3 ENTER 4 y^x	81
2.25 ENTER 4 y^x	25.6283
− Γ• πx	173.9534
64 ÷	2.718

It is frequently necessary to calculate the deflection or sag of a beam when a load is imposed. There are a number of deflection formulas for different load configurations. We will illustrate the deflection formula for deflection at midspan for a concentrated load applied at midspan. The equation is

$$y = \frac{FL^3}{48\ EI} \tag{7}$$

where

  y = deflection, in
  F = force, lb
  L = length of beam, in
  E = the elastic modulus of the material ($3E7$ lb/in^2 for steel)
  I = the moment of inertia, in^4

EXAMPLE 13. Calculate the deflection at the center of a span for a steel beam from the following data: F = 6000 lb, L = 20 ft = 240 in, E = $3E7$ lb/in^2, I = 227 in^4. (See Fig. 3.4.)

Given: a 6000-lb load is applied at the center of a 20-ft beam: I = 227 in^4, E = $3E7$ psi.

Required: Find the deflection at midspan.

$$y = \frac{FL^3}{48\ EI} = \frac{(6000\ \text{lb})(240\ \text{in})^3}{(48)(3E7\ \text{psi})(227\ \text{in}^4)} = 0.254\ \text{in}$$

Calculator routine for algebraic notation:

ENTER	PRESS	DISPLAY
6000	× (	
240	y^x	
3	) +	
48	+	
30	EE	
6	+	
227	=	2.537 −1

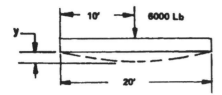

FIG. 3.4

Calculator routine for reverse Polish notation:

PRESS	DISPLAY
⌐ DISP (SC 3)	
240 ENTER 3 y^x	1.382 E7
6000 ×	8.294 E10
48 ÷	1.728 E9
3 E7 ÷	5.76 E1
227 ÷	2.537 E −1

Another structural member that is frequently analyzed is the column. There are a number of different column formulas based on different materials and codes. We illustrate the formula used by the American Institute for Steel Construction for intermediate length columns (Bassin et al., 1969):

$$\frac{F}{A} = \frac{S_y}{N}\left[1 - \frac{(Kl/r)^2}{2C_c^2}\right] \tag{8}$$

where

$\dfrac{F}{A}$ = allowable stress, force per unit area, lb/in^2

S_y = yield strength of the material, lb/in^2

N = factor of safety

K = a factor dependent on the end retainment of the column

L = length of the column, in

C_c = elasticity factor of the material

r = radius of gyration of the column, in

EXAMPLE 14. Calculate the allowable stress for a column from the following data: $S_y = 36000$ lb/in^2, N = 1.88, K = 1, L = 170 in, r = 2.35 in, C_c = 170.

$$\frac{F}{A} = \frac{S_y}{N}\left[1 - \frac{(Kl/r)^2}{2C_c^2}\right]$$

$$= \frac{36,000}{1.88}\left[1 - \frac{(170/2.35)^2}{2 \times 170^2}\right] = 17,415 \text{ lb/in}^2$$

Calculator routine for algebraic notation:

ENTER	PRESS	DISPLAY
36,000	÷	
1.88	× (	
1	− (	
170	÷	
2.35	) x^2 ÷	
2	÷ (	
170	x^2)) =	17,415

Note the use of two levels of parentheses.

It is best to start within the innermost parentheses for complex equations using reverse Polish notation:

Calculator routine for reverse Polish notation:

PRESS	DISPLAY
⌐ DISP (FX 4)	
170 <u>ENTER</u> 2.35 ÷	72,3404
⌐ x^2	5,233.13
2 + 170 ⌐ x^2 ÷	0.905
+/− 1 +	0.9095
36000 ×	32,740
1.88 ÷	17,415

Another application of powers is in the design of springs. Consider the following equation, which is used to determine the number of active coils in a compression spring (Spotts, 1971):

$$N = \frac{\delta d^4}{6.956E\text{-}7PD^3} \tag{9}$$

where

N = number of active coils
δ = spring deflection, in
d = the wire diameter, in
P = the load force, lb
D = the mean coil diameter, in

and 6.956E-7 is in units of in²/lb

EXAMPLE 15. Calculate the number of active coils for the spring shown in Fig. 3.5 with the following given data: δ = 3 in, d = 0.375 in, P = 480 lb, D = 1.8 in.

Given: A spring carries a compressive load of 480 lb. The mean diameter D is 1.8 in. Wire diameter d is 0.375 in.

Required: Find the number of active coils N for a deflection δ of 3 in.

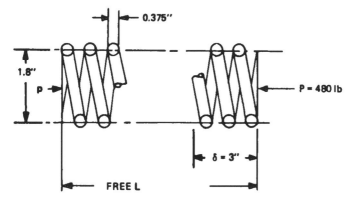

FIG. 3.5

$$N = \frac{\delta d^4}{6.956\text{E-}7 \ PD^3}$$

$$= \frac{(3 \text{ in})(0.375 \text{ in})^4}{6.956\text{E-}7 \text{ in}^2/\text{lb} \times 480 \text{ lb} \times (1.8 \text{ in})^3} = 30.5$$

Calculator routine for algebraic notation:

PRESS	DISPLAY
3 × (0.375 y^x 4)	
+ 6.956 EE (−) 7	
+ 480 ÷	
(1.8 y^x 3) =	3.0466 E01 or 30.46

Calculator routine for reverse Polish notation:

PRESS	DISPLAY
0.375 ENTER 4 y^x	0.0198
3 ×	0.0593
6.957 E7 +/− +	85,275.5
480 ÷	177.6573
1.8 ENTER 3 y^x +	30.46

This topic should not be concluded without a discussion of negative powers and roots. A negative power or root of a number means simply the reciprocal of the positive power or root of that number. For example, $4^{-3} = 1/4^3 = 0.015625$ and is solved on the calculator as follows:

Calculator routine for algebraic notation:

PRESS	DISPLAY
2nd FIX 4	
4 y^x (−) 3 =	0.0156

Calculator routine for reverse Polish notation:

PRESS	DISPLAY
4 ENTER 3 +/− y^x	0.0156

Another example follows:

$-\sqrt[3]{27} = 1/\sqrt[3]{27} = 0.333333$, is solved on the calculator as follows:

Calculator routine for algebraic notation:

PRESS	DISPLAY
(−) 3 2nd $\sqrt[x]{y}$ 27 =	0.3333

Calculator routine for reverse Polish notation:

PRESS	DISPLAY
27 ENTER	
3 +/− $\frac{1}{x}$	−0.3333
y^x	0.3333

V. The Pythagorean Theorem

One of the most commonly used principles in the solution of technical and scientific problems is the Pythagorean theorem, which involves the three sides of a right triangle. It states that in a right triangle, the square of the hypotenuse is equal to the sum of the squares of the other two sides. Algebraically this can be written:

$$C^2 = A^2 + B^2 \tag{10}$$

where

C is the hypotenuse and A and B are the other two sides.

With your calculator this problem can be solved quickly and easily.

EXAMPLE 16.   In a right triangle A = 16, B = 12. Solve for C.

$$C = \sqrt{16^2 + 12^2} = 20$$

Calculator routine for algebraic notation:

PRESS	DISPLAY
$\sqrt{}$ (16 x^2 + 12 x^2) =	20

Calculator routine for reverse Polish notation:

PRESS	DISPLAY
16 $\rceil$ x^2	256
12 $\rceil$ x^2 + $\sqrt{x}$	20

If the hypotenuse and one other side are known, the equation is:

$$B = \sqrt{C^2 - A^2} \tag{11}$$

EXAMPLE 17. In a right traingle C = 525, A = 350. Solve for B.

$$B = \sqrt{525^2 - 350^2} = 391.3$$

Calculator routine for algebraic notation:

PRESS	DISPLAY
$\sqrt{}$ (525 x^2 −	
350 x^2) =	391.311

Calculator routine for reverse Polish notation:

PRESS	DISPLAY
525 $\rceil$ x^2	275,625
350 $\rceil$ x^2	122,500
− $\sqrt{x}$	391.311

The Pythagorean theorem is used in navigation, surveying, structural design, electrical problems, and many other areas.

The variety of problems in this chapter should give the student an appreciation of the many uses of these functions in problem solving.

Exercise 3.1

Solve for x using the reciprocal function:

3.1-1 $\dfrac{1}{200} + \dfrac{1}{300} = \dfrac{1}{x}$

3.1-2 $\dfrac{1}{6.15} + \dfrac{1}{8.3} + = \dfrac{1}{x}$

3.1-3 $\dfrac{1}{60,000} + \dfrac{1}{20,000} - \dfrac{1}{30,000} = \dfrac{1}{x}$

3.1-4 $\dfrac{1}{235} + \dfrac{1}{750} + \dfrac{1}{50} = \dfrac{1}{x}$

3.1-5 $\dfrac{1}{25} + \dfrac{1}{75} - \dfrac{1}{80} = \dfrac{1}{x}$

3.1-6 For the circuit shown in Fig. 3.6, calculate (a) the net resistance, (b) I for each branch, (c) the IR drop for each branch.

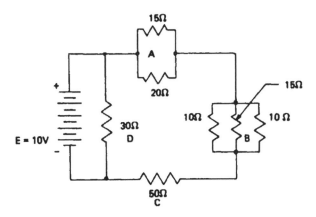

FIG. 3.6

3.1-7 A projector lens has a focal length of 12 cm. Solve for the object distance when the image distance is 6 cm.

Exercise 3.2

3.2-1 Solve for the roots of the following quadratic equation:

$x^2 + 459x - 14,217 = 0$

3.2-2 Solve for the roots of the equation:

$16x^2 + 1100x - 6,270 = 0$

3.2-3 Calculate the moment of inertia for a rectangular beam where

$b = 2$ in, $h = 6$ in.

3.2-4   Calculate the moment of inertia for a 1.25-in-diameter shaft.

3.2-5   Calculate the moment of inertia for a tube having an O.D. of 2.125 in and an I.D. of 1.857 in.

3.2-6   Calculate the deflection at midspan of a 10-ft steel beam when a 5,000-lb load is imposed at midspan. I = 109.7 in^4, E = 3E7 lb/in^2.

3.2-7   Calculate the deflection at midspan of a 12-ft wooden beam with a cross section of b = 6 in and h = 12 in when a 3,000-lb load is imposed at midspan. E = 1.76E6 lb/in^2.

3.2-8   Calculate F/A for a column when the following data are given: S_y = 36,000 lb/in^2, N = 1.89, L = 288 in, K = 1, r = 3.08 in, C_e = 126.1.

3.2-9   Calculate the number of active coils for a spring when δ = 5.0 in, d = 0.3065 in, D = 4.1 in, P = 160 lb.

3.2-10   Evaluate 0.55^{-3}.

Exercise 3.3

Given two sides in Fig. 3.7, solve for the third, rounding off the answers to the same degree of accuracy as the data given.

FIG. 3.7

3.3-1   A = 25 in, B = 15 in
3.3-2   A = 312 ft, B = 120 ft
3.3-3   A = 41.5 in, B = 27.3 in
3.3-4   A = 515 cm, B = 250 cm
3.3-5   A = 735 m, B = 430 m
3.3-6   C = 45 in, A = 25 in
3.3-7   C = 150 ft, A = 85 ft
3.3-8   C = 375 m, A = 125 m
3.3-9   C = 515 mi, A = 420 mi
3.3-10   C = 1000 km, A = 625 km

4 · Trigonometric Functions

I. Explanation of Basic Trigonometric Functions

The solution of many types of technical problems involves the use of trigonometry, the literal meaning of which is *triangle measurement*. In a right triangle there is always an exact relationship between the three sides for any given angle relationship.

Referring to Fig. 4.1, the functions are defined as follows:

$$\text{sine (sin) } \theta \quad = \frac{\text{opposite}}{\text{hypotenuse}}$$

$$\text{cosine (cos) } \theta \quad = \frac{\text{adjacent}}{\text{hypotenuse}}$$

$$\text{tangent (tan) } \theta \quad = \frac{\text{opposite}}{\text{adjacent}}$$

$$\text{secant (sec) } \theta \quad = \frac{\text{hypotenuse}}{\text{adjacent}} = \frac{1}{\cos}$$

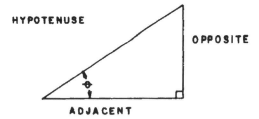

FIG. 4.1

$$\text{cosecant (csc) } \theta = \frac{\text{hypotenuse}}{\text{opposite}} = \frac{1}{\sin}$$

$$\text{cotangent (cot) } \theta = \frac{\text{adjacent}}{\text{opposite}} = \frac{1}{\tan}$$

Up until recent years the solution of trigonometry problems involved referring to tables for the values of the different functions. All the calculators we are considering here have trigonometric function keys for sine, cosine, and tangent. The trigonometric functions can be set directly for any angle entered in either decimal degrees or radians. (Radian measurement of angles will be discussed later.) Be sure to set the *DEG-RAD* (D-R) switch in the desired position before beginning trigonometric operations.

To enter a trigonometric function, enter the angle, followed by the desired function.

EXAMPLE 1. In a right triangle the angle θ is 36.55° and the length of the hypotenuse is 89.35 in. Solve for the length of the opposite side. Stating the definition of the sine function and transposing:

Opposite = hypotenuse × sin θ = 89.35 × sin 36.55° = 53.21 in (See Fig. 4.1.)

Calculator routine for algebraic notation:

PRESS	DISPLAY
89.35 × <u>SIN</u> 36.55 =	53.2101

Calculator routine for reverse Polish notation:

PRESS	DISPLAY
89.35 <u>ENTER</u>	
36.55 <u>SIN</u> ×	53.2101

A note of caution should be mentioned here. Because the calculator takes an instant to calculate the value of the trigonometric function, you should wait until this value appears in the display before pressing the key for the next operation or entry.

EXAMPLE 2. In a right triangle the angle is 25.67° and the length of the opposite side is 10.56 in. Solve for the length of the hypotenuse. Again we transpose the definition of the sine function:

$$\text{Hypotenuse} = \frac{\text{opposite}}{\sin \theta} = \frac{10.56 \text{ in}}{\sin 25.67°} = 24.38 \text{ in}$$

Calculator routine for algebraic notation:

PRESS	DISPLAY
10.56 ÷ <u>SIN</u>	
25.67 =	24.377

Calculator routine for reverse Polish notation:

PRESS	DISPLAY
10.56 <u>ENTER</u>	
25.67 <u>SIN</u> ÷	24.377

EXAMPLE 3. In a right triangle the angle is 36.5° and the length of the hypotenuse is 14.35 cm. Solve for the length of the adjacent side. Transpose the definition for the cosine function:

$$\text{Adjacent} = \text{hypotenuse} \times \cos \theta$$
$$= 14.35 \text{ cm} \times \cos 36.5° = 11.54 \text{ cm}$$

Calculator routine for algebraic notation:

PRESS	DISPLAY
14.35 × <u>COS</u> 36.5 =	11.535

Calculator routine for reverse Polish notation:

PRESS	DISPLAY
14.35 <u>ENTER</u>	
36.5 <u>COS</u> ×	11.535

EXAMPLE 4. In a right triangle the angle is 20.47° and the length of the adjacent side is 35.25 ft. Solve for the length of the opposite side. Transpose the definition of the tangent function:

Opposite = adjacent × tan θ = 35.25 ft × tan 20.47° = 13.16 ft

Calculator routine for algebraic notation:

PRESS	DISPLAY
35.25 ×	
<u>TAN</u> 20.47 =	13.16

Calculator routine for reverse Polish notation:

PRESS	DISPLAY
35.25 <u>ENTER</u>	
20.47 <u>TAN</u> ×	13.16

II. Inverse Trigonometric Functions

When the ratio of the lengths of two sides of a right triangle is known, it is frequently necessary to determine the corresponding angle. In mathematical symbols this relationship is often expressed as arcsin, arccos, arctan or $\sin^{-1}$, $\cos^{-1}$, $\tan^{-1}$. In any case the meaning is as follows: the angle whose sine is . . . .

Here the procedures for all three calculators vary somewhat. For the TI-68, press "INV" followed by the function involved. For the FX-7000GA press "SHIFT" followed by "SIN^{-1}", "COS^{-1}", or "TAN^{-1}". For the HP-32SII, press "↰" followed by "ASIN", "ACOS", or "ATAN".

EXAMPLE 5. Arcsin 0.563 = 34.26°.

Calculator routine for the TI-68:

PRESS	DISPLAY
<u>INV</u> <u>SIN</u> 0.563 =	34.26

Calculator routine for the FX-7000GA:

PRESS	DISPLAY
<u>SHIFT</u> <u>SIN</u>$^{-1}$ 0.563 <u>EXE</u>	34.26

Calculator routine for the HP-32SII:

PRESS	DISPLAY
0.563 **⌐** <u>ASIN</u> 34.26	

EXAMPLE 6.   An airplane pilot flying a plane with a cruising speed of 150 mi/hr wishes to plot a course which is directly east. He will encounter a wind blowing at 30 mi/hr out of the north. What will be his heading to fly the desired course? (See Fig. 4.2.)

Given: An airplane cruises at 150 mi/hr; a north wind blows at 30 mi/hr.

Required: Find the heading to fly straight east.

$$\theta = \arcsin \frac{30}{150} = 11.54°$$

Calculator routine for the TI-68:

PRESS	DISPLAY
<u>INV SIN</u> (30 ÷ 150) = 11.54	

Calculator routine for the FX-7000GA:

PRESS	DISPLAY
<u>SHIFT</u> <u>SIN</u>$^{-1}$ (30 ÷ 150) = 11.54	

Calculator routine for the HP-32SII:

PRESS	DISPLAY
30 <u>ENTER</u> 150 ÷ **⌐** <u>ASIN</u> 11.54	

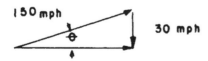

FIG. 4.2

III. Functions In Radians

Although we are accustomed to measuring angles in degrees, we must remember that the degree is an arbitrary unit. For many scientific equations it is necessary to evaluate angles with the radian, which is a natural unit. The radian is the measure of an angle whose arc along the circumference of a circle is just equal to the radius. (See Fig. 4.3.) It follows that there are 2π radians in a circle.

The distance that an object travels in a circular path is given by the equation

$$s = r\theta \tag{1}$$

where

    s = distance traveled, linear units
    r = radius of the circle, linear units
    θ = angle traveled, radians

EXAMPLE 7. The mean radius of the earth's orbit around the sun is 92,957,000 mi. How far does the earth travel in orbit in 3 months? (See Fig. 4.4.)

Given: The mean radius of the earth's orbit is 92,957,000 mi.

Required: Find the distance traveled in orbit in 3 months.

$$s = r\theta = \frac{9.257\text{E}7 \text{ mi} \times \pi}{2} = 1.46\text{E}8 \text{ mi}$$

A disk cam is a disk with an irregular profile designed to produce a desired sequence of motion to a follower which presses against it as it rotates. One of the most familiar examples is the application in internal

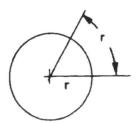

FIG. 4.3

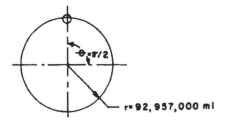

r=92,957,000 mi

FIG. 4.4

combustion engines, where cams control the opening and closing of the valves. For high-speed operation, cam contours must be plotted from mathematical equations. One motion used in cam design is called cycloidal motion. Positions on the cam contour are plotted from the equation:

$$s = \frac{\theta}{\beta} L - \frac{L}{2\pi} \sin \frac{2\pi\theta}{\beta} \qquad (2)$$

where

    s = distance cam follower moves, in
    L = total lift of the cam, in
    β = total radians of motion
    θ = angle for plotted position, radians

Before performing calculations in the radian mode it will be necessary to explain the methods of setting the radian mode on the various calculators. For the TI-68 press "3rd DRG." This key sets the calculator for degrees, radians, or grads. Note the small figure at the top of the display. If it is not "R," press "3rd DRG" again until it appears. For the FX-7000GA press "MODE 5." For the HP-32SII press "⌐ MODES." A menu appears:

    DG RD GR

Press the key in the top row below "RD." "RAD" in small figures appears at the top of the display.

EXAMPLE 8. Solve for s if L = 1, θ = π/6, β = π. (See Fig. 4.5.)

Given: A cycloidal cam has a lift of 1 in per 180° (π radians).

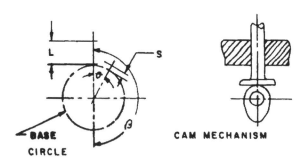

FIG. 4.5

Required: Calculate the rise when cam has turned 30° ($\pi/6$ radians).

$$s = \frac{\theta}{\beta} L - \frac{L}{2\pi} \sin \frac{2\pi\theta}{\beta}$$

$$= \frac{\pi/6 \times 1}{\pi} - \frac{1}{2\pi} \sin \frac{2\pi}{\pi} \frac{\pi}{6} = 0.0288 \text{ in}$$

Calculator routine for the TI-68:

PRESS	DISPLAY
3rd <u>DRG</u> (Until R appears)	
6 x^{-1} − (2 × <u>2nd</u> π) x^{-1}	
× <u>SIN</u> (<u>2nd</u> π ÷ 3) =	0.0288

Calculator routine for the FX-7000GA:

PRESS	DISPLAY
<u>MODE</u> 5	Radians
6 x^{-1} − (2 × <u>SHIFT</u> π) x^{-1}	
× <u>SIN</u> (<u>SHIFT</u> π ÷ 3) <u>EXE</u>	0.0288

Calculator routine for the HP-32SII:

PRESS	DISPLAY
⌐ <u>MODES</u> (RD)	
⌐ π <u>ENTER</u> 3 ÷ <u>SIN</u> 2 ÷	
⌐ π ÷ <u>+/−</u> 6 $\frac{1}{x}$ +	0.0288

IV. Radian-Degree Conversion

Sometimes it is necessary to make conversions between radians and degrees. Remember that π radians $= 180°$. To convert radians to degrees, multiply by $180/\pi$. To convert degrees to radians, multiply by $\pi/180$. All three calculators have routines for making this conversion, but this method is simpler.

Continuing with radian application problems, we present a problem involving radians and convert the answer to degrees. The formula for calculating the angular deflection at the end of a beam resting on two supports when a load is applied at midspan is

$$\theta = \frac{FL^2}{16EI} \qquad\qquad (3)$$

where

θ = angular deflection, rad
L = length, in
F = load, lb
E = modulus of elasticity, 3E7 lb/in^2 for steel
I = moment of inertia, in^4

EXAMPLE 9. Calculate the angular deflection at the end of a 10-in-long by 1.25-in-diameter steel shaft when a 1000-lb load is applied at the center. Convert the answer to degrees. (See Fig. 4.6.)

Given: A 1.25-in-diameter shaft with a span of 10 in between bearings has a 1000-lb load applied at midspan.

Required: Find the angular deflection at B_1.

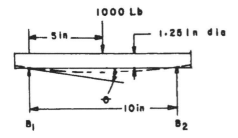

FIG. 4.6

$$I = \frac{\pi d^4}{64} = \frac{\pi (1.25)^4}{64} = 0.1198 \text{ in}^4$$

$$\theta = \frac{FL^2}{16EI} = \frac{(1000 \text{ lb})(10 \text{ in})^2}{(16)(30E6 \text{ lb/in}^2)(0.1198 \text{ in}^4)} = 1.739E\text{-}3 \text{ rad}$$

$$= 0.0996°$$

Calculator routine for the TI-68:

PRESS	DISPLAY
2nd π × (1.25 y^x 4)	
÷ 64 = STO ALPHA I =	0.1198
1000 × 10 x^2 ÷ 16 ÷ 30 E6	
÷ ALPHA I =	0.0017 radians
× 180 ÷ 2nd π =	0.0996 degrees

Calculator routine for the FX-7000GA:

PRESS	DISPLAY
MODE 5 EXE	
SHIFT π × 1.25 X^Y 4 ÷ 64	
EXE → ALPHA A EXE	0.1198
1000 × 10 x^2 ÷ 16 ÷	
30 EXP 6 ÷ ALPHA A EXE	0.0017 radians
× 180 ÷ SHIFT π =	0.0996 degrees

Calculator routine for the HP-32SII:

PRESS	DISPLAY
◄┐ DISP (EN 3)	
┌► π ENTER 1.25 ENTER 4 y^x	
× 64 ÷ STO A	
1000 ENTER 10 ◄┐ x^2 ×	
16 ÷ 30 E6 ÷ RCL A ÷	1.738 E-3 radians
180 × ┌► π ÷	99.6 E-3
◄┐ DISP (FX 4)	0.0996 degrees

V. Decimal Degree to Degree-Minute-Second Conversions

The calculator ordinarily deals with decimal degrees when solving trigonometric problems. All three calculators have routines for making conversions between decimal degrees and degrees, minutes, and seconds.

For the TI-68 the conversion from degrees, minutes, and seconds to decimal degrees is performed as follows: press "DMS" after degrees, minutes, and seconds. Press "2nd DD = ." To convert decimal degrees to degrees, minutes, and seconds, press "INV 2nd DD = ."

For the FX-7000GA the conversion from degrees, minutes, and seconds to decimal degrees is performed as follows: press ",,," after degrees, minutes, and seconds. Press "EXE." To convert decimal degrees to degrees, minutes, and seconds, (accomplished only after a previous calculation) press "SHIFT ,,,."

For the HP-32SII the conversion from degrees, minutes, and seconds to decimal degrees is performed as follows: "← → HR." Press the key in the top row below "HR." To convert decimal degrees to degrees, minutes, and seconds, press "→HMS."

Press the key in the top row below "HR." To convert decimal degrees to degrees, minutes, and seconds, press "→HMS."

You must remember in which mode you entered the angle. There is no way that you can tell from looking at the display. The angle 13°16′35″ will be entered in the display as 13.1635.

EXAMPLE 10. Find the value of sin 13°16′35″.

Calculator routine for the TI-68:

PRESS	DISPLAY
13 DMS	
16 DMS	
35 DMS	
2nd DD =	13.2764
STO ALPHA A =	
SIN ALPHA A =	0.2296

Calculator routine for the FX-7000GA:

PRESS	DISPLAY
SIN 13 ,,, 16 ,,, 35 ,,, EXE	0.2296

Calculator routine for the HP-32SII:

PRESS	DISPLAY
13.1635 ⌐ →HR	13.276
SIN	0.2296

EXAMPLE 11. Find the inverse tangent in degrees, minutes, seconds for a right triangle where the opposite side is 256.00 ft and the adjacent side is 395.00 ft.

Calculator routine for the TI-68:

PRESS	DISPLAY
256 ÷ 395 =	0.6481
STO ALPHA T =	
INV TAN ALPHA T =	32.9473 decimal deg.
INV 2nd DD =	32°56′50.36″

Calculator routine for the FX-7000GA:

PRESS	DISPLAY
SHIFT TAN⁻¹	
(256 ÷ 395) EXE	32.9473 decimal deg.
SHIFT ,,,	32°56′50.36″

Calculator routine for the HP-32SII:

PRESS	DISPLAY
256 ENTER 395 ÷ ⌐ ATAN	32.9473 decimal deg.
⌐ →HMS	32.5650 (deg. min. sec.)

VI. Vector Problems

In engineering a number of quantities are analyzed by trigonometric methods. Trigonometry is used to deal not only with distances but also with forces, velocities, accelerations, and electrical relationships. In this type of analysis any quantity which can be represented by a line in both

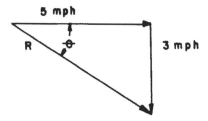

FIG. 4.7

magnitude and direction is referred to as a factor. If two or more vectors are involved, the quantity equivalent to the net effect of the other vectors is called the resultant.

EXAMPLE 12. A boat heads across a stream at a speed of 5 mi/hr. The stream flows at 3 mi/hr. What is the boat's actual speed and direction? (See Fig. 4.7.)

Given: Boat heads across stream at 5 mph.

Required: Find boat's actual speed and heading if stream flows at 3 mph.

$$R = \sqrt{(5 \text{ mph})^2 + (3 \text{ mph})^2} = 5.83 \text{ mph}$$

$$\theta = \arctan \frac{3}{5} = 30.96°$$

EXAMPLE 13. A weight of 500 lb is suspended from a cable. It is deflected to the right by a horizontal force which causes the cable to make an angle of 20° with the vertical. Calculate the tension in the cable and the magnitude of the horizontal force. The vectors can be arranged in a triangle, placing each force parallel to its position in the first diagram. (See Fig. 4.8.)

Given: A 500-lb force suspended from a cable *AC* is deflected to the right 20°.

Required: Solve for forces BC and AC.

$$BC = 500 \text{ lb} \times \tan 20° = 182 \text{ lb}$$

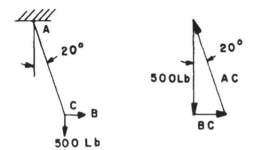

FIG. 4.8

$$AC = \frac{500 \text{ lb}}{\cos 20^\circ} = 532 \text{ lb}$$

For alternating current, the relationship between current and voltage does not follow the simple relationship stated in Ohm's law. For alternating current the current reverses through each cycle with the magnitude of the current varying as a sine wave. Two additional factors must be considered: capacitance and inductance. The voltage across a capacitor lags the current by 90°, and the voltage across an inductance leads the current by 90°. The effective resistance across any part of the circuit is called the reactance and its symbol is X. Capacitive reactance is denoted by X_C. Inductive reactance is denoted by X_L. The total impedance Z for an RC circuit is expressed by the formula (Harris and Hemmerling, 1969; Malmstadt et al., 1963):

$$Z = \sqrt{R^2 + X_C^2} \tag{4}$$

The relationship is represented by the impedance vector diagram shown in Fig. 4.9. The phase angle between applied potential and current is determined by

$$\tan \theta = \frac{X_C}{R} \tag{5}$$

In a similar manner, the impedance for a circuit containing a resistance and an inductance is expressed by the formula

$$Z = \sqrt{R^2 + X_L^2} \tag{6}$$

The relationship is shown in Fig. 4.10.

The phase angle between applied potential and current is determined

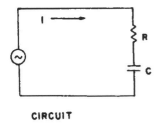

CIRCUIT

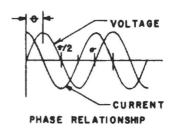

PHASE RELATIONSHIP

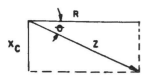

IMPEDANCE DIAGRAM

FIG. 4.9 Series RC circuit and impedance diagram.

by

$$\tan \theta = \frac{X_L}{R} \tag{7}$$

If both an inductance and a capacitance occur in the same circuit, the total impedance is expressed by the formula

$$Z = \sqrt{R^2 + (X_L - X_C)^2} \tag{8}$$

$$\tan \theta = \frac{X_L - X_C}{R} \tag{9}$$

Capacitances and inductances are often used in combination in order to reduce the phase angle θ. Ohm's law for ac circuits containing capacitances and inductances in series becomes

$$I = \frac{E}{Z} \tag{10}$$

As current leads or lags voltage in RC or RL circuits, it is advantageous to keep the phase angle as small as possible. The cosine of the phase angle is called the power factor. For alternating current:

$$P = EI \cos \theta \tag{11}$$

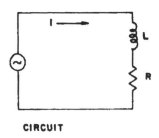

CIRCUIT

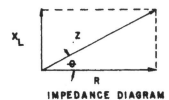

IMPEDANCE DIAGRAM

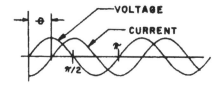

PHASE RELATIONSHIP

FIG. 4.10 Series LR circuit and impedance diagram.

EXAMPLE 14. In the RC circuit shown in Fig. 4.11, R = 10 kΩ and X_C = 12.5 kΩ. Solve for the impedance Z and the phase angle θ between the applied potential and the current.

Given: A series RC circuit with a resistance of 10 kΩ and a capacitive reactance of 12.5 kΩ.

Required: Solve for Z and θ.

$$Z = \sqrt{R^2 + X_C^2} = \sqrt{(10 \text{ k}\Omega)^2 + (12.5 \text{ k}\Omega)^2} = 16 \text{ k}\Omega$$

$$\theta = \arctan \frac{X_C}{R} = \arctan \frac{12.5}{10} = 51.34°$$

EXAMPLE 15. In the series RLC circuit shown in Fig. 4.12, R = 50 Ω, X_L = 75 Ω, X_C = 40 Ω. Calculate the impedance and power factor.

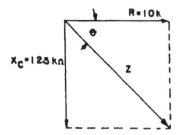

FIG. 4.11

Given: In a series RLC circuit R = 50 Ω; X_L = 75 Ω; X_C = 40 Ω;

Required: Calculate impedance and power factor.

$$Z = \sqrt{R^2 + (X_L - X_C)^2} = \sqrt{50^2 + (75 - 40)^2} = 61.03 \; \Omega$$

$$\tan \theta = \frac{X_L - X_C}{R} = \frac{75 - 40}{50}$$

$$\theta = 34.99°$$

Power factor = $\cos \theta$ = 0.819

Because all vector problems do not lend themselves to right triangle solutions, other procedures are needed for the solution of oblique triangles. These will be covered in a later section.

VII. Rectangular-Polar Coordinate Conversions

Two systems are used for plotting points. In the rectangular coordinate system two axes are arranged at right angles. The X axis is horizontal

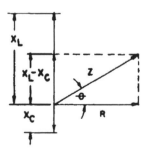

FIG. 4.12

and the Y axis is vertical. Values of X are positive when plotted to the right of the Y axis and negative when plotted to the left of the Y axis. Values of Y are positive when plotted above the X axis and negative when plotted below the X axis. In the polar coordinate system the length of the radius R and the angle θ of the radius measured counterclockwise from the X axis are used. The conversion from one system to the other is simple in principle.

EXAMPLE 16. Convert the rectangular coordinates x = 4, y = 3 to polar coordinates R, θ.

$$R = \sqrt{4^2 + 3^2} = 5$$

$$\theta = \arctan \frac{3}{4} = 36.87°$$

EXAMPLE 17. Convert the polar coordinates R = 10, θ = 35.5° to rectangular coordinates x, y.

$$x = 10 \cos 35.5° = 8.141$$

$$y = 10 \sin 35.5° = 5.807$$

All three of the calculators have routines for making these conversions. However, making the conversion from polar to rectangular is simpler when using the equations shown above. We will illustrate the procedures for converting from rectangular to polar, but the equations shown in example 16 are just about as easy.

Calculator routine for algebraic notation:

PRESS	DISPLAY
10 × COS 35.5 =	8.141
10 × SIN 35.5 =	5.807

Calculator routine for reverse Polish notation:

PRESS	DISPLAY
10 ENTER 35.5 COS ×	8.141
10 ENTER 35.5 SIN ×	5.807

The routine for converting rectangular to polar coordinates is as follows:

EXAMPLE 18.   Convert $x = 4$, $y = 3$ to polar coordinates.

Calculator routine for the TI-68:

PRESS	DISPLAY
(4 $\vdots$ 3) <u>INV</u>	
<u>2nd P→R</u> =	5 radians
	36.87 degrees

Calculator routine for the FX-7000GA:

PRESS	DISPLAY
<u>SHIFT</u> <u>POL</u> (4	
<u>SHIFT</u> $\vdots$ 3) <u>EXE</u>	5 radians
<u>ALPHA</u> J <u>EXE</u>	36.87 degrees

Calculator routine for the HP-32SII:

PRESS	DISPLAY
3 <u>ENTER</u> 4	
⌐ →θ, r	5 radians
x ↘ y	36.87 degrees

This procedure saves time when you are determining the resultant of three or more vectors. Basically, the solution of this type of problem involves the following steps:

1. Evaluate the projection of each vector on the x and y axes.
2. Calculate the summation of all the x projections.
3. Calculate the summation of all the y projections.
4. Calculate the resultant from

$$R = \sqrt{(\Sigma x)^2 + (\Sigma y)^2}$$

5. Calculate θ from

$$\theta = \arctan \frac{\Sigma y}{\Sigma x}$$

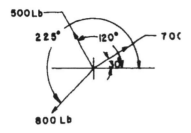

FIG. 4.13

EXAMPLE 19. Determine the magnitude and direction of the resultant for the force system illustrated in Fig. 4.13.

Given: Force system shown in Fig. 4.13.

Required: Solve for magnitude and direction of resultant.

F, lb	Angle	x, lb	y, lb
700	30°	606.2	350
500	120°	−250	433
800	225°	−565.7	−565.7
	Σ	−209.5	217.3

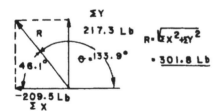

Solution for Example 19.

VIII. Sine and Cosine Law Problems

For the solution of oblique triangles there are four combinations of known parts from which a solution may be obtained. These combinations are (Washington, 1970):

Case 1. Two angles and one side
Case 2. Two sides and the angle opposite one of them

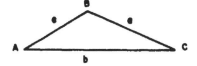

LAW OF SINES

$$\frac{a}{\sin A} = \frac{b}{\sin B} = \frac{c}{\sin C}$$

LAW OF COSINES

$$c^2 = a^2 + b^2 - 2ab \cos C$$

FIG. 4.14 Sine and cosine laws.

Case 3. Two sides and the included angle
Case 4. Three sides

Let us consider the oblique triangle illustrated in Fig. 4.14. The angles are designated A, B, and C. The sides opposite the angles are designated a, b, and c. Cases 1 and 2 can be covered by the law of sines, which states that the ratio of any side to the sine of the angle opposite it is equal to the ratio of any other side to the sine of the angle opposite it:

$$\frac{a}{\sin A} = \frac{b}{\sin B} = \frac{c}{\sin C} \qquad (12)$$

EXAMPLE 20. In Fig. 4.15 let angle A = 30°, angle C = 25°, side a = 10 in. Solve for side b.

Given: Triangle with data shown.

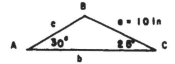

FIG. 4.15

Required: Solve for side b.

$$B = 180° - 30° - 25° = 125°$$

By the law of sines:

$$\frac{10 \text{ in}}{\sin 30°} = \frac{b}{\sin 125°}$$

$$b = 16.38 \text{ in}$$

Cases 3 and 4 can be covered by the law of cosines, which states that when any two sides and the included angle are known, the square of the third side is equal to the sum of the squares of the known sides minus 2 times the product of the known sides times the cosine of the included angle (see Fig. 4.14):

$$c^2 = a^2 + b^2 - 2ab \cos C \tag{13}$$

or

$$c = \sqrt{a^2 + b^2 - 2ab \cos C} \tag{14}$$

EXAMPLE 21.   An airplane pilot wishes to set a course to a point which is due east. His ground speed is to be 600 mi/hr and he will encounter a 40 mi/hr wind blowing out of the northwest. What will be his air speed and heading? (See Fig. 4.16.) Note that in solving this problem good use can be made of the memory capability of the calculator.

Given: Desired ground speed for plane is 600 mph. A 40-mph wind is blowing from the northwest.

Required: Calculate air speed and heading.

From law of cosines:

$$\text{air speed} = \sqrt{(600)^2 + (40)^2 - 2 \times 600 \times 40 \times \cos 45°}$$

$$= 572 \text{ mph}$$

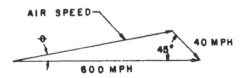

FIG. 4.16

From law of sines:

$$\frac{572}{\sin 45°} = \frac{40}{\sin \theta}$$

$$\theta = 2.83°$$

Calculator routine for algebraic notation:

PRESS	DISPLAY
600 <u>STO</u> <u>ALPHA</u> A =	
40 <u>STO</u> <u>ALPHA</u> B =	
$\sqrt{}$ (ALPHA A x^2 +	
<u>ALPHA</u> B x^2 − 2 ×	
<u>ALPHA</u> A × <u>ALPHA</u> B ×	
<u>COS</u> 45) =	572.4
x^{-1} × 40 × <u>SIN</u> 45 =	0.0494
<u>INV</u> <u>SIN</u> 0.0494 =	2.83 degrees

Calculator routine for reverse Polish notation:

PRESS	DISPLAY
600 <u>STO</u> A	
40 <u>STO</u> B	
<u>RCL</u> A ⅂ x^2	
<u>RCL</u> B ⅂ x^2 +	
2 <u>ENTER</u> <u>RCL</u> A × <u>RCL</u> B ×	
45 <u>COS</u> × − $\sqrt{x}$	572.4
$\frac{1}{x}$ 40 × 45 <u>SIN</u> ×	0.0494
⅂ <u>ASIN</u>	2.83 degrees

For case 4, where three sides of a triangle are known and an angle between two of the sides is to be calculated, the law of cosines takes this form:

$$\cos A = \frac{b^2 + c^2 - a^2}{2bc} \tag{15}$$

EXAMPLE 22. Determine angle A for the triangle shown in Fig. 4.17.

Given: Triangle with three known sides as shown.

Required: Solve for angle A.

$$\cos A = \frac{b^2 + c^2 - a^2}{2bc}$$

$$= \frac{(4.800)^2 + (5.900)^2 - (3.600)^2}{2 \times 5.900 \times 4.800} = 0.7925494$$

$$A = 37.5756°$$

Calculator routine for algebraic notation:

PRESS	DISPLAY
3.6 <u>STO</u> <u>ALPHA</u> A =	
5.9 <u>STO</u> <u>ALPHA</u> B =	
4.8 <u>STO</u> <u>ALPHA</u> C =	
<u>INV</u> <u>COS</u> ((<u>ALPHA</u> B x² +	
<u>ALPHA</u> C x² −	
<u>ALPHA</u> A x²) ÷ 2	
÷ <u>ALPHA</u> B + <u>ALPHA</u> C) =	37.57

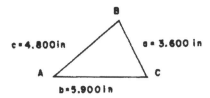

c = 4.800 in   a = 3.600 in

b = 5.900 in

FIG. 4.17

Calculator routine for reverse Polish notation:

PRESS	DISPLAY
3.6 <u>STO</u> A	
4.8 <u>STO</u> B	
5.9 <u>STO</u> C	
<u>RCL</u> B ⌐ x^2	
<u>RCL</u> C ⌐ x^2 +	
<u>RCL</u> A ⌐ x^2 −	
2 <u>ENTER</u> <u>RCL</u> C ×	
<u>RCL</u> B × $\dfrac{1}{x}$ ×	0.7925
⌐ <u>ACOS</u>	37.576

In this chapter we have tried to give some idea of the scope of trigonometry applications in engineering and of the methods of setting up calculator solutions to trigonometry problems. This knowledge will be of use as you encounter more complex problems in other courses.

Exercise 4.1

4.1-1   Solve for x in Fig. 4.18a.

4.1-2   Solve for x in Fig. 4.18b.

4.1-3   Solve for x in Fig. 4.18c.

4.1-4   Solve for x in Fig. 4.18d.

4.1-5   Solve for x in Fig. 4.18e.

4.1-6   Solve for θ in Fig. 4.18f.

4.1-7   Solve for θ in Fig. 4.18g.

4.1-8   Solve for θ in Fig. 4.18h.

4.1-9   Solve for θ in Fig. 4.18j.

4.1-10   Solve for θ in Fig. 4.18k.

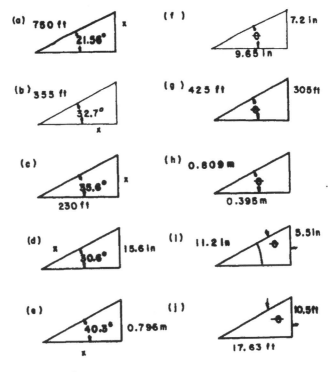

FIG. 4.18

Exercise 4.2

Convert degrees to radians:

4.2-1   26.55°
4.2-2   95.6°
4.2-3   120°

Convert radians to degrees:

4.2-4 $\pi/6$ radians
4.2-5 $\pi/3$ radians
4.2-6 $2\pi/3$ radians
4.2-7 $4\pi/5$ radians
4.2-8   Convert 47°26′10″ to decimal degrees.
4.2-9   Convert 37.565° to degrees, minutes, seconds.
4.2-10   For a cycloidal cam, calculate the value of s when the following data are given: $\theta = 15°$, $\beta = 120°$, $L = 1$ in.

4.2-11   A 1-in-diameter steel shaft rests on two bearings 8 in apart. A load of 500 lb is applied at the center of the span. Calculate the angular deflection of the shaft at the bearings.

$I = \pi d^4/64$, $E = 3E7$ lb/in^2.

Convert answer to **degrees, minutes, seconds**.

Exercise 4.3

Convert to polar coordinates R, θ:

4.3-1   x = 25, y = 15
4.3-2   x = − 10, y = 30
4.3-3   x = −25, y = −13.5
4.3-4   x = 130, y = −75

Convert to rectangular coordinates x, y:

4.3-5   R = 10, θ = 35.5°
4.3-6   R = 25.5, θ = 120°
4.3-7   R = 103, θ = 210°
4.3-8   R = 17.5, θ = 315°
4.3-9   Calculate the magnitude and direction of the force system shown in Fig. 4.19.

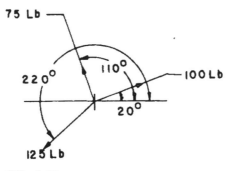

FIG. 4.19

4.3-10   In a series LRC circuit R = 50 Ω, X_L = 270 Ω, X_C = 75 Ω. Solve for impedance Z and phase angle between applied potential and current.

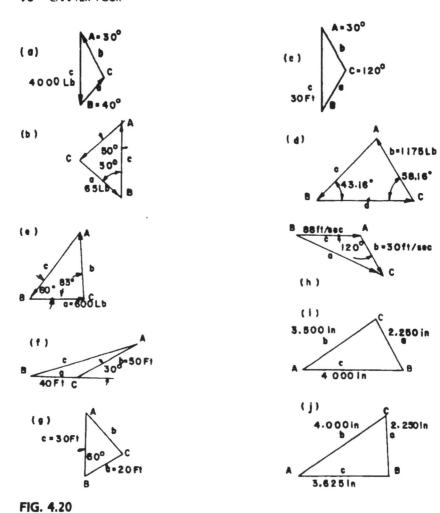

FIG. 4.20

Exercise 4.4

4.4-1   Solve for all the unknowns in Fig. 4.20a.
4.4-2   Solve for all the unknowns in Fig. 4.20b.
4.4-3   Solve for all the unknowns in Fig. 4.20c.
4.4-4   Solve for all the unknowns in Fig. 4.20d.
4.4-5   Solve for all the unknowns in Fig. 4.20e.
4.4-6   Solve for all the unknowns in Fig. 4.20f.

4.4-7   Solve for all the unknowns in Fig. 4.20g.

4.4-8   Solve for all the unknowns in Fig. 4.20h.

4.4-9   Solve for all the unknowns in Fig. 4.20i.

4.4-10   Solve for all the unknowns in Fig. 4.20j.

5 · Physics Problems and Formula Rearrangement

I. Principles of Formula Rearrangement

Consider a formula as an abbreviated way of expressing a series of mathematical operations. One should always think of the meaning of the formula rather than merely memorizing a set of symbols. Consider the following formula (Bassin et al., 1969), which is basic in stress calculations:

$$\sigma = \frac{F}{A} \tag{1}$$

where

σ = stress, lb/in^2 or psi
F = force, lb
A = area, in^2

This equation gives a relationship between three variables, any two of which may be known. In determining the strength of a steel specimen in the tensile testing machine, the known variables are force and area. In designing a structural member to carry a given load, force and allowable stress are the known variables. In determining the safe load that a structural member of known cross section will support, allowable stress and area are the two known variables. In each of the last two examples, it is necessary to rearrange the formula so that the unknown variable is on the left side.

$$A = \frac{F}{\sigma} \tag{2}$$

becomes

$$F = \sigma A \tag{3}$$

This, of course, is a basic algebraic operation. Two fundamental rules to remember are

1. A term that is a multiplier becomes a divisor when transferred to the other side, and vice versa.
2. A term that is an addend becomes a subtrahend when transferred to the other side, and vice versa.

It should be noted how the units in the answer are arrived at. Let us look again at the stress equation:

$$\sigma \ (\text{lb/in}^2) = \frac{F \ (\text{lb})}{A \ (\text{in}^2)}$$

Or, let us look at a more complex equation (Harris and Hemmerling, 1972):

$$E \ (\text{lb/in}^2) = \frac{F \ (\text{lb}) \ L \ (\text{in})}{A \ (\text{in}^2) \ e \ (\text{in})}$$

It can be seen that the units on the right side of the equation cancel out to give the units on the other side.

II. Motion Problems

In order to provide more experience with formula arrangement, we will present some of the basic motion equations of physics. It is not our intention to go deeply into theory here; this will be found in the study of physics. Our purpose is to acquaint the student with some of the practical applications of formula rearrangement which affect our daily lives.

The quantities we will deal with are:

1. Distance, represented by s, may be expressed in ft, mi, or km
2. Velocity, represented by v, may be expressed in mi/hr, ft/sec, m/sec, or km/hr
3. Time, represented by t, may be in sec, min, hr
4. Acceleration, which is the rate of change of velocity, represented by the symbol a, usually expressed in ft/sec^2, ft/(sec)(sec), or m/sec^2, m/(sec)(sec).

The basic formulas for the condition where there is no initial velocity are (Harris and Hemmerling, 1972):

$$s = vt \tag{4}$$

$$s = \frac{1}{2} at^2 \tag{5}$$

$$v = at \tag{6}$$

$$v^2 = 2as \tag{7}$$

It is essential that compatible units be applied in the equations. For example, in equation (4) if s is expressed in feet, v should be expressed in ft/sec. In general practice mi/hr is converted to ft/sec when motion equations are involved. To convert 60 mi/hr to ft/sec:

$$\frac{60 \text{ mi/hr} \times 5280 \text{ ft/mi}}{3600 \text{ sec/hr}} = 88 \text{ ft/sec}$$

This is a good ratio to remember. To convert mi/hr to ft/sec, multiply by 88/60. To convert ft/sec to mi/hr, multiply by 60/88.

Let us proceed with some example problems.

EXAMPLE 1.   A car is traveling at 50 mi/hr when the driver applies the brakes. If he brings the car to a stop in 100 ft, what is his rate of negative acceleration? Looking through the motion formulas, we find that equation (7), $v^2 = 2as$, contains the variables involved. However, the unknown in this case is a. Placing a on the left, the equation becomes

$$a = \frac{v^2}{2s}$$

Next we convert 50 mi/hr to ft/sec.

$$50 \text{ mi/hr} \times \frac{88 \text{ ft/sec}}{60 \text{ mi/hr}} = 73.33 \text{ ft/sec}$$

$$a = \frac{v^2}{2s} = \frac{(73.33 \text{ ft/sec})^2}{2 \times 100 \text{ ft}} = 26.87 \text{ ft/sec}^2$$

EXAMPLE 2. What was the time required for the car in the previous example to stop? In this case equation (5), $s = (1/2)at^2$ applies. Rearranging to solve for t, the formula becomes

$$t = \sqrt{\frac{2s}{a}} = \sqrt{\frac{2 \times 100 \text{ ft}}{26.87 \text{ ft/sec}^2}} = 2.728 \text{ sec}$$

When initial velocities are considered, the basic motion equations become

$$s = v_1 t + \frac{1}{2} at^2 \tag{8}$$

$$v_2 = v_1 + at \tag{9}$$

$$v_2^2 = v_1^2 + 2as \tag{10}$$

where v_1 is initial velocity and v_2 is final velocity.

EXAMPLE 3. How far will a car travel while accelerating from 30 to 60 mi/hr if the rate of acceleration is 25 ft/sec^2? In this case the applicable formula is (10). Rearranged to solve for s, it becomes

$$s = \frac{v_2^2 - v_1^2}{2a} = \frac{(88 \text{ ft/sec})^2 - (44 \text{ ft/sec})^2}{2 \times 25 \text{ ft/sec}^2} = 116.2 \text{ ft}$$

Calculator routine for algebraic notation:

ENTER	PRESS	DISPLAY
88	($\underline{x}^2$ −	
44	$\underline{x}^2$) ÷	
2	+	
25	=	116.16 (116.2)

Calculator routine for reverse Polish notation:

PRESS	DISPLAY
88 ⬏ x^2	7744
44 ⬏ x^2 −	5802
2 ÷ 25 +	116.16

III. Motion Problems Involving g

The generally accepted values for acceleration due to the earth's gravitational pull at sea level are 32.17 ft/sec^2 for the English system of measurement and 9.81 m/sec^2 for the SI system. For problems involving the earth's gravity these values are substituted as g in place of a in the motion equations previously shown:

$$s = 1/2gt^2 \tag{11}$$

$$v = gt \tag{12}$$

$$v^2 = 2gs \tag{13}$$

$$s = v_1t + 1/2gt^2 \tag{14}$$

$$v_2 = v_1 + gt \tag{15}$$

$$v_2 = v_1^2 + 2gs \tag{16}$$

EXAMPLE 4. How long will it take for an object in free fall to fall 100 ft? Neglect the effect of air friction. In this case equation (11) applies. Rearranging this formula to solve for t:

$$t = \sqrt{\frac{2s}{g}} = \sqrt{\frac{2 \times 100 \text{ ft}}{32.17 \text{ ft/sec}^2}} = 2.493 \text{ sec}$$

EXAMPLE 5. An object is in free fall at 100 ft/sec. How long will it take to reach a velocity of 150 ft/sec? In this case equation (15) applies. Solving for t, this becomes

$$t = \frac{v_2 - v_1}{g} = \frac{150 \text{ ft/sec} - 100 \text{ ft/sec}}{32.17 \text{ ft/sec}^2} = 1.554 \text{ sec}$$

IV. Ratio and Proportion

A ratio is a comparison of two like quantities by division. Any measurement made is the ratio of the measured magnitude to that of an accepted standard of measurement. When we say that an object is 5 ft long, we are saying that the length of the object is five times that of an accepted standard, the foot. Other examples of ratios are density (weight/volume), specific gravity (density of object/density of water), and pressure (force/area).

A statement of equality between two ratios is called a proportion. A typical proportion is

$$\frac{x}{6} = \frac{4}{8}$$

The algebraic solution follows the rule that the product of the means is equal to the product of the extremes:

$$8x = (4)(6)$$

$$x = \frac{(4)(6)}{8} = 3$$

On the calculator this amounts to a chain multiplication, a process with which you are already familiar.

A direct proportion is one in which two variables increase or decrease at a uniform rate. A good example is the relationship of velocity and distance.

EXAMPLE 6.   If the average speed of a car is 55 mi/hr, how far will it travel in 7.3 hr? The proportion equation is

$$\frac{x \text{ mi}}{7.3 \text{ hr}} = \frac{55 \text{ mi}}{1 \text{ hr}}$$

Therefore

$$x = \frac{(7.3 \text{ hr})(55 \text{ mi})}{1 \text{ hr}} = 401.5 \text{ mi}$$

In an inverse proportion one variable decreases as the other increases. A good example is the relationship of velocity and time.

EXAMPLE 7.   If a jet plane flies at the rate of 600 mi/hr for 2.5 hr, how long will it take a slower plane traveling at 275 mi/hr to cover the same distance? The proportion is

$$\frac{275 \text{ mi/hr}}{600 \text{ mi/hr}} = \frac{2.5 \text{ hr}}{x \text{ hr}}$$

$$x = \frac{(2.5 \text{ hr})(600 \text{ mi/hr})}{275 \text{ mi/hr}} = 5.45 \text{ hr}$$

There are many applications of ratio and proportion in engineering. One that is frequently used is the solution of similar triangles.

EXAMPLE 8.   In the triangle shown in Fig. 5.1, AC, BC, and AD are known. The length of DE is required.

Given: A triangle with the dimensions shown.

Required: Solve for ED.

$$\frac{ED}{95} = \frac{100}{150}$$

$$ED = \frac{100 \text{ ft} \times 95 \text{ ft}}{150 \text{ ft}} = 63.33 \text{ ft}$$

Another application is the solution of speed changes in gear and belt drives. Here the principle is that the speeds of two shafts are inversely proportional to the numbers of teeth in the gears on the two shafts or to the diameters of the pulleys on the two shafts.

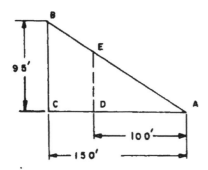

FIG. 5.1

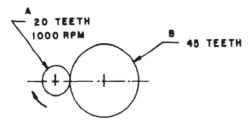

FIG. 5.2

EXAMPLE 9. As shown in Fig. 5.2, a shaft with a 20-tooth gear rotating at 1000 rpm drives another shaft with a 45-tooth gear. What is the speed of the second shaft?

Given: Gear A, with 20 teeth, rotates at 1000 rpm. Gear B has 45 teeth.

Required: Calculate rpm of B.

$$\frac{N_A}{N_B} = \frac{(rpm)_B}{(rpm)_A}$$

$$\frac{20}{45} = \frac{x}{1000}$$

$$x = \frac{20 \text{ teeth} \times 1000 \text{ rpm}}{45 \text{ teeth}} = 444.4 \text{ rpm}$$

V. Other Formula Rearrangement Problems

Of particular interest to students of chemistry and physics are the gas law problems. As a gas is a form of matter which has neither a definite shape nor a definite volume, three variables are involved: volume, pressure, and temperature. The first gas law, Boyle's law, deals with the relationship between volume and pressure. It states that when the temperature remains constant, volume varies inversely with absolute pressure (Harris and Hemmerling, 1972):

$$\frac{P_1}{V_2} = \frac{P_2}{V_1} \tag{17}$$

where

V_1 = initial volume

V_2 = second volume

P_1 = initial absolute pressure

P_2 = second absolute pressure

Absolute pressure must include standard barometric pressure, which is equivalent to 76 cm of mercury or 14.7 lb/in². If gage pressure is in the given data, standard barometric pressure must be added to gage pressure to give absolute pressure.

EXAMPLE 10. A gas cylinder contains 4 ft³ of gas at 100 psi gage pressure. If the temperature is not changed, what will be the gage pressure if the gas is transferred to a 10 ft³ volume container?

Solution: First, let us identify the known quantities:

V_1 = 4 ft³

P_1 = 100 psi + 14.7 psi = 114.7 psi absolute (psia)

V_2 = 10 ft³

Applying Boyle's law, equation (17), and rearranging the formula:

$$P_2 = \frac{P_1V_1}{V_2} = \frac{114.7 \text{ psia} \times 4 \text{ ft}^3}{10 \text{ ft}^3} = 45.88 \text{ psia}$$

Subtracting atmospheric pressure:

45.88 psia − 14.7 psi = 31.18 psi gage

The second gas law, Charles' law, deals with the temperature volume relationship. It states that when the pressure remains constant, the volume of a gas varies directly in proportion to its absolute temperature (Harris and Hemmerling, 1972):

$$\frac{V_1}{T_1} = \frac{V_2}{T_2} \tag{18}$$

where

V_1 = initial volume

V_2 = second volume

T_1 = initial absolute temperature

T_2 = second absolute temperature

Extensive research has established that the lowest temperature attainable is 273° below 0° Celsius. This is referred to as zero degrees

Kelvin, or 0 K. If one is using the Farenheit scale, absolute zero is 460° below 0°F. This is referred to as 0° Rankine.

EXAMPLE 11. The volume of a gas is 5 liters at 25°C. If it is compressed to a volume of 2 liters without a change of pressure, what will be its final Celsius temperature?

Solution: Identifying the known quantities:

V_1 = 5 liter

V_2 = 2 liter

T_1 = 25°C + 273° = 298 K

Applying Charles' law, equation (18), and rearranging the formula:

$$T_2 = \frac{V_2 T_1}{V_1} = \frac{5\rlap{/}{L} \times 298 \text{ K}}{2\rlap{/}{L}} = 745 \text{ K}$$

Converting to Celsius:

745 K − 273° = 472°C

In actual practice, all three variables—volume, temperature, and pressure—interact. If we combine Boyle's law and Charles' law, we obtain the universal gas law (Harris and Hemmerling, 1972):

$$\frac{P_1 V_1}{T_1} = \frac{P_2 V_2}{T_2} \tag{19}$$

EXAMPLE 12. In a laboratory experiment, it is desired to find the density of air. A 1-liter container equipped with a stopcock is weighed on a precision balance. The air is removed with a vacuum pump, and the container is again weighed. The net difference in weight is 1.16 g. Room temperature is 23°C. The barometer reading is 74 cm of mercury. It is desired to express the density at standard conditions of temperature and pressure, 273 K and 76 cm of mercury.

Solution: First we solve for the equivalent volume for standard conditions of temperature and pressure, STP:

V_1 = 1 liter

T_1 = 23°C + 273° = 296 K

T_2 = 273 K

$P_1 = 74$ cm

$P_2 = 76$ cm

Applying the universal gas law, equation (19), and rearranging the formula:

$$V_2 = \frac{P_1 V_1 T_2}{T_1 P_2} = \frac{74 \cancel{cm} \times 1 \text{ liter} \times 273 \text{ K}}{296 \text{ K} \times 76 \cancel{cm}} = 0.898 \text{ liter}$$

Solving for density at STP:

$$\text{Density} = \frac{\text{Mass}}{\text{Volume}} = \frac{1.16 \text{ g}}{0.898 \text{ liter}} = 1.29 \text{ g/liter}$$

EXAMPLE 13. A typical application of the gas laws is to be found in the anlaysis of the function of an internal combustion engine. An internal combustion engine has a compression ratio of 7.5 to 1. The ambient temperature (local temperature) is 90°F. The temperature in the cylinder at the top of the compression stroke is 450°F. Solve for the absolute pressure in psi at the top of the compression stroke.

$V_1 = 7.5$

$V_2 = 1$

$T_1 = 90°F + 460° = 550°R$

$T_2 = 450°F + 460° = 910°R$

$P_1 = 14.7$ psia

Applying the universal gas law and rearranging the formula:

$$P_2 = \frac{P_1 V_1 T_2}{T_1 V_2} = \frac{14.7 \text{ psi} \times 7.5 \times 910°R}{550°R \times 1} = 182.4 \text{ psia}$$

Finally, there is the application of the gas laws to chemistry. It has been established that 1 g molecular weight (1 mole) of any gas occupies a volume of 22.4 liters at STP (273 K and 76 cm of mercury). Knowing this relationship, it is possible to calculate the mass of any volume of gas or the volume of any mass of gas, provided that its true chemical formula and molecular weight are known (Sorum, 1969).

EXAMPLE 14. Determine the weight of 100 liters of CO_2 at a temperature of 85°C and a pressure of 50 cm of mercury.

Solution: First, calculate the equivalent volume at STP:

$V_1 = 100$ liters

$P_1 = 50$ cm

$P_2 = 76$ cm

$T_1 = 85°C + 273° = 350$ K

$T_2 = 273$ K

Applying the universal gas law and rearranging the formula:

$$V_2 = \frac{P_1 V_1 T_2}{T_1 P_2} = \frac{50 \text{ cm} \times 100 \text{ liters} \times 273 \text{ K}}{358 \text{ K} \times 76 \text{ cm}} = 50.17 \text{ liters}$$

Solution: Calculate the number of moles:

$$\frac{50.17 \text{ liters}}{22.4 \text{ liters/mole}} = 2.24 \text{ moles}$$

Calculate the weight of 1 mole of CO_2:

12 g/mole C + 2 × 16 g/mole O = 44 g/mole CO_2

Calculate the total weight:

2.24 mole × 44 g/mole = 98.56 g

The examples set forth in this chapter should help give the student a better understanding of the application of formula rearrangement in problem solving.

Exercise 5.1

Solve the following problems, rearranging the formula and stating the answer in the proper units:

5.1-1 Solve for A if s = F/A, where s = 30,000 psi, F = 1000 lb.

5.1-2 Solve for e if E = FL/Ae, where E = 3E7 psi, F = 2500 lb, L = 50 in, A = 0.75 in^2.

5.1-3 Solve for a if s = (1/2)at^2, where t = 5 sec, s = 1000 ft.

5.1-4 Solve for a if $v_2 = v_1 + at$, where $v_1 = 44$ ft/sec, $v_2 = 75$ ft/sec, t = 5 sec.

5.1-5 How long will it take a car traveling at 60 mi/hr to slow to 20 mi/hr if a = 20 ft/sec^2?

5.1-6   A ball is thrown vertically at a velocity of 70 ft/sec. What is its velocity after 4 sec (g should be applied as a negative value)?

5.1-7   How long will it take an object in free fall to fall 150 ft?

5.1-8   The muzzle velocity of a rifle is 2600 ft/sec. If the barrel is 30 in long, what is the acceleration of the bullet?

Exercise 5.2

5.2-1   30 mi/hr = 44 ft/sec. How many ft/sec is 100 mi/hr?

5.2-2   If a car travels 100 mi on 4.5 gal of fuel, how far will it go on 10 gal?

5.2-3   The weight of a sphere varies as the cube of the radius. If a 3-in radius sphere weighs 100 lb, what is the weight of a 6-in radius sphere of the same material?

5.2-4   The wind pressure on a wall varies directly as the area A and as the square of the wind velocity v. If the force on a 12 × 18-ft wall is 120 lb when v = 15 mi/hr, what is the force on a 10 × 20-ft wall when v = 25 mi/hr?

5.2-5   According to Boyle's law of gases pressure varies inversely with volume:

$$\frac{P_1}{P_2} = \frac{V_2}{V_1}$$

If V_1 is 60 in³, P_1 is 150 psi, and V_2 is 20 in³, solve for P_2.

5.2-6   In Fig. 5.3 is shown a V-belt drive. The drive sheave has a 4-in diameter and rotates at 1750 rpm. What diameter driven sheave will be required to give a speed of 600 rpm?

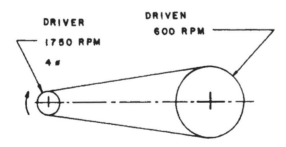

FIG. 5.3

5.2-7   If a 36 × 96-in sheet of #10 gage sheet steel weighs 135 lb, how much does a 48 × 144-in sheet weigh?

5.2-8 In Fig. 5.4 solve for the length of AD, using the principle of similar triangles. (The length AB will first have to be found by using the Pythagorean theorem.)

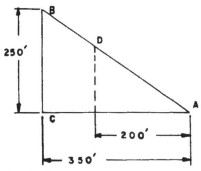

FIG. 5.4

Exercise 5.3

5.3-1 A compressed air tank with a volume of 4 ft³ contains air at atmospheric pressure. How much additional air at atmospheric pressure must be pumped in to give a gage pressure of 100 psi? Use Boyle's law and subtract the volume of the tank from the volume of the air pumped in.

5.3-2 How much air at atmospheric pressure must be pumped into a tire of 1600-in³ volume to raise the gage pressure from 24 to 32 psi? Consider the volume of the tire as constant. Apply Boyle's law twice, converting atmospheric air to air under pressure; then solve for the two volumes of atmospheric air.

5.3-3 A pressure tank for a home water system has a volume of 150 gal. If the tank is initially filled with air at atmospheric pressure, what will be the resultant gage pressure after 100 gal of water is pumped into the tank? Use Boyle's law.

5.3-4 In a laboratory experiment the air is evacuated from a 550-cm³ container. The barometer reading is 73 cm of mercury and the temperature in the room is 25°C. The established density of air at STP is 1.293 gm/liter. What mass of air should be evacuated from the container? Use the universal gas law.

5.3-5 In a diesel engine the compression ratio is 16 to 1. Ambient air at 70°F and 14.7 psi is drawn into the cylinders. The temperature at the top of the compression stroke is 1000°F. What is the pressure at the top of the compression stroke?

5.3-6 Determine the mass of 1000 liters of methane gas (CH_4) at a temperature of 25°C and a pressure of 125 cm of mercury.

5.3-7 What volume will 150 g of NO gas occupy at a temperature of 20°C and a pressure of 100 cm of mercury?

.

6 · Calculation of Volumes, Centroids, and Center of Gravity

Many engineering calculations are based on the topics covered in this chapter. The weights of materials in structures, the flow of fluids, the calculation of earth to be moved in cuts and fills, and many other calculations are based on the accurate determination of volumes. The determination of centroids is an important part of stress calculations as well as a key to the calculation of volume of irregular figures. Calculation of the center of gravity of bodies is an important step in determining the behavior of these bodies when they are in motion.

Although many of these calculations involve the simple application of formulas, some include a number of steps. The main purpose of this chapter is to provide training in organizing such problems and in making advantageous use of the memory capacity of the calculator in their solution.

I. Volumes and Areas of Simple Geometric Solids

We have already discussed problems involving some simple volumes such as calculation of the volume of a cylinder. In this section we deal

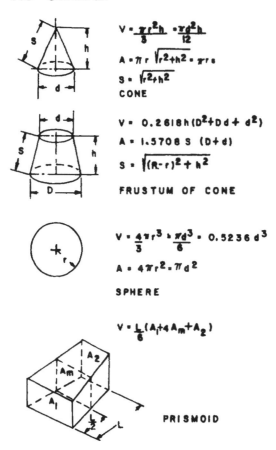

$$V = \frac{\pi r^2 h}{3} = \frac{\pi d^2 h}{12}$$

$$A = \pi r \sqrt{r^2+h^2} = \pi r s$$

$$s = \sqrt{r^2+h^2}$$

CONE

$$V = 0.2618 h (D^2+Dd+d^2)$$

$$A = 1.5708 \, s \, (D+d)$$

$$s = \sqrt{(R-r)^2+h^2}$$

FRUSTUM OF CONE

$$V = \frac{4 \pi r^3}{3} = \frac{\pi d^3}{6} = 0.5236 \, d^3$$

$$A = 4 \pi r^2 = \pi d^2$$

SPHERE

$$V = \frac{L}{6}(A_1+4A_m+A_2)$$

PRISMOID

FIG. 6.1 Volume formulas.

with other area and volume formulas that are commonly used (see Fig. 6.1) (Oberg and Jones, 1964).

One will probably have more occasion to calculate the volume and area of a frustum of a cone than that of a cone. All kinds of hoppers, air ducts, etc., involve this geometrical shape.

EXAMPLE 1. Calculate the volume of a conical hopper which is of 40-in diameter at the upper end and 10-in diameter at the lower end with a height of 35 in.

Given: A hopper of the dimensions shown in Fig. 6.2.

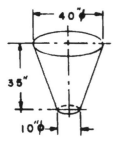

FIG. 6.2

Required: Calculate the volume.

$V = 0.2618h(D^2 + Dd + d^2)$

$= 0.2618 \times 35\ [(40)^2 + (40 \times 10) + (10)^2] = 19,240\ \text{in}^3$

There are some interesting problems involving the sphere.

EXAMPLE 2. A spherical gas tank is to be designed. The desired volume is 5000 ft^3. Calculate the diameter and surface area of this tank.

Solution: Referring to Fig. 6.1, we find the formula for the volume of a sphere:

$V = 0.5236d^3$

Rearranging the formula and solving for d:

$$d = \sqrt[3]{\frac{V}{0.5236,}} = \sqrt[3]{\frac{5000\ \text{ft}^3}{0.5236}} = 21.216\ \text{ft}$$

Calculating the area:

$A = \pi d^2 = \pi(21.216)^2 = 1414\ \text{ft}^2$

It is often necessary to determine the volume of the contents of a horizontal drum by measuring the depth of the liquid in the drum. This involves calculating the area of a circular segment. A segment is the portion of a circle under a chord of the circle. The area of a circular segment is determined by first calculating the area of a sector (see Chapter 2) and then subtracting the area of the triangular portion of the sector from its total area. The area of the segment multiplied by the length of the drum gives the volume. As can be seen, a number of steps are involved. There area also two different procedures, depending on whether the depth of the fluid is less than or greater than the radius of the drum. In solving this problem, good use of the calculator's memory registers can be made.

EXAMPLE 3. A 24-in diameter drum of 40-in length lies in a horizontal position. Calculate the volume in gallons when the measured depth of the liquid is (a) 8 in, (b) 16 in.

Given: A 24-in diameter drum 40 in long lies in a horizontal position.

Required: Calculate the volume in gallons when depth of fluid is 8 in, as shown in Fig. 6.3.

Calculator routine for algebraic notation:

PRESS	DISPLAY	COMMENT
4 + 12 =		
STO ALPHA A =	0.333	COS α
INV COS ALPHA A =	70.529	α
STO ALPHA A2 =		
× 2 = STO ALPHA A1 =	141.058	2α
12 × SIN ALPHA A2 =	11.314	BC
× 4 = STO ALPHA B =	45.255	Area $\triangle$ AOB
ALPHA A1 × 2nd π × 12		
x^2 + 360 =	177.258	Area sector AOB
− ALPHA B =	132.003	Area segment
× 40 + 231 =	22.858	Vol., gal.

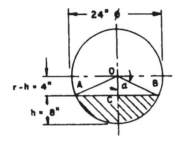

FIG. 6.3

Calculator routine for reverse Polish notation:

PRESS	DISPLAY	COMMENT
4 ENTER 12 ÷		
⌐ ACOS STO A	70.529	α
2 × STO B	141.058	2α
12 ENTER RCL A SIN ×	11.314	BC
4 × STO C	45.255	Area Δ AOB
RCL B 360 ÷		
⌐ π ×		
12 ⌐ x² ×	177.26	Area sector
RCL C − STO D	132.003	Area segment
40 × 231 ÷	22.858	Vol., gal.

Required: Calculate the volume in gallons when depth of fluid is 16 in, as shown in Fig. 6.4.

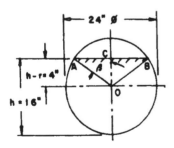

FIG. 6.4

Calculator routine for algebraic notation:

PRESS	DISPLAY	COMMENT
4 ÷ 12 = STO ALPHA B =	0.333	COS B
INV COS ALPHA B =	70.529	B
STO ALPHA B1 = × 2 =	141.058	2 B
STO ALPHA B2 =		
12 × SIN ALPHA B1 =	11.314	BC
× 4 = STO ALPHA C =	45.255	Area Δ AOB
ALPHA B2 × 2nd π × 12		
x^2 + 360 =	177.258	Area sector
− ALPHA C = STO ALPHA D =	132.003	Area segment
2nd π × 12 x^2 −		
ALPHA D =	320.386	Eff. area
× 40 ÷ 231 =	55.478	Vol., gal.

Calculator routine for reverse Polish notation:

PRESS	DISPLAY	COMMENT
⌈ π 12 ⌉ x^2 ×		
RCL D −	320.386	Eff. area
40 × 231 ÷	55.478	Vol., gal.

II. Volume of a Prismoid

The prismoidal formula is a general formula by which the volume of any prism, pyramid, or frustum of a pyramid can be found. As can be seen, a prismoid is a tapered prismlike figure bounded by parallel planes at opposite ends which have different areas. The figure is commonly used by civil engineers and architects in cut-and-fill calculations for earth removal. It is also used in other engineering calculations which will not be explained here. The planes may be rectangular, triangular, or trapezoidal. To determine the dimensions of the center plane, it is necessary to take the average of the corresponding dimensions of the end planes.

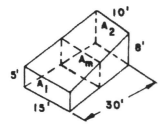

FIG. 6.5

The area of the center plane is not the average of the areas of the end planes.

EXAMPLE 4. Calculate the volume of the prismoid shown in Fig. 6.5.

Calculator routine for algebraic notation:

PRESS	DISPLAY	COMMENT
(15 + 10) ÷ 2 =	12.5	Base Am
STO ALPHA B =		
(5 + 8) ÷ 2 =	6.5	Height Am
STO ALPHA H =		
30 ÷ 6 × ((15 × 5) +		
(4 × ALPHA B × ALPHA H) +		
(8 × 10)) =	2400.	Vol.

Calculator routine for reverse Polish notation:

PRESS	DISPLAY	COMMENT
15 ENTER 10 + 2 ÷ STO A	12.5	Base Am
5 ENTER 8 + 2 ÷ STO B	6.5	Height Am
15 ENTER 5 ×		
4 ENTER RCL A ×		
RCL B × + 8 ENTER 10 × +		
30 × 6 +	2400	Vol.

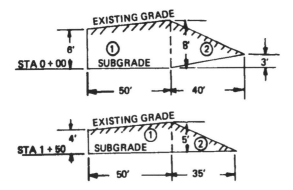

FIG. 6.6

EXAMPLE 5. In Fig. 6.6 are shown two sections through a site to be excavated. The sections are 150 ft apart. Calculate the amount of earth to be removed. The volume can be divided into two prismoids, 1 and 2. The dimensions shown in the sections are those of the two end planes of the prismoids. In the equations the use of the calculator memories is indicated. It is recommended that in the solution of this type of problem, you record all the intermediate answers, as shown. This will enable you to check them if necessary.

Required: Calculate the volume between the sections shown (in yd³).

Calculator routine for algebraic notation:

PRESS	DISPLAY	REMARKS
(6 + 4) ÷ 2 =	5	(b1)
STO ALPHA A =		
(5 + 8) ÷ 2 =	6.5	(b2)
STO ALPHA B =		
150 ÷ 6 × (((6 + 8)		
÷ 2 × 50) +		
(4 × ((ALPHA A + ALPHA B)		
÷ 2 × 50) + ((4 + 5) ÷ 2		
× 50)) =	43,123 ft.³	(vol. 1)

PRESS	DISPLAY	REMARKS
STO ALPHA C =		
(35 + 40) ÷ 2 =	37.5	(h1)
STO ALPHA D =		
150 ÷ 6 × ((8 × 40 ÷ 2)		
+ (4 × ALPHA B × ALPHA D		
÷ 2) + (5 × 35 ÷ 2)) =	18,375 ft.^{3}	(vol. 2)
+ ALPHA C ÷ 27 =	2,278 yd.^{3}	(total vol.)

Calculator routine for reverse Polish notation:

PRESS	DISPLAY	REMARKS
6 ENTER 4 + 2 ÷	5.0	(b1)
STO A		
5 ENTER 8 + 2 ÷	6.5	(b2)
STO B		
6 ENTER 8 + 2 ÷ 50 ×		
RCL A RCL B + 2 ÷		
4 × 50 × +		
4 ENTER 5 + 2 ÷ 50 × +		
150 ENTER 6 ÷ × STO C	43,125 ft.^{3}	(vol. 1)
35 ENTER 40 + 2 ÷	37.5	(h1)
STO D		
8 ENTER 40 × 2 ÷		
RCL B RCL D × 4 × 2 ÷ +		
5 ENTER 35 × 2 ÷ +		
150 ENTER 6 ÷ ×	18,375 ft.^{3}	(vol. 2)
RCL C + 27 ÷	2,278 yd.^{3}	(total vol.)

III. Centroids

The center of gravity of a body is the point at which all the weight in the body can be balanced. Strictly speaking, there is no center of gravity of an area, for an area does not have weight. Because of its relation to the center of gravity, the center of an area is referred to as the centroid. Finding the location of the centroid of the cross section of a beam is an important step in the stress analysis of a beam.

In Fig. 6.7 is shown a tabulation of the values of centroids for simple areas. Always assume that the lower lefthand corner of the figure lies

SHAPE	AREA	$\bar{x}$	$\bar{y}$	
RECTANGLE ·	bh	$\frac{b}{2}$	$\frac{h}{2}$	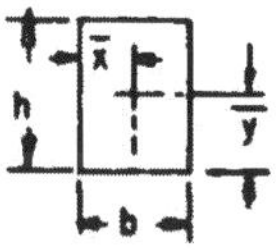
TRIANGLE	$\frac{bh}{2}$	$\frac{b}{3}$	$\frac{h}{3}$	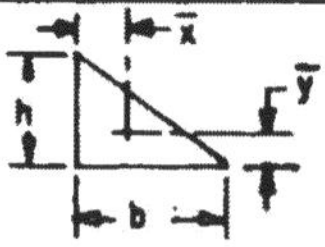
CIRCLE	$\frac{\pi d^2}{4}$	$\frac{d}{2}$	$\frac{d}{2}$	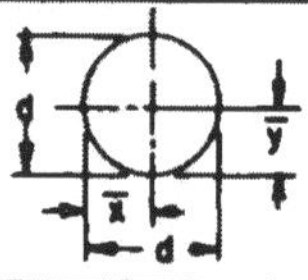
SEMICIRCLE	$\frac{\pi d^2}{8}$	$\frac{d}{2}$	$\frac{4r}{3\pi}$ (0.425r)	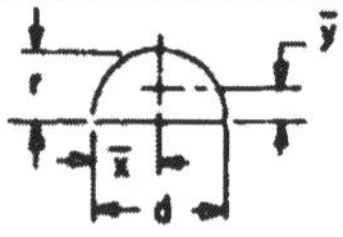
QUADRANT	$\frac{\pi d^2}{16}$	$0.425r$	$0.425r$	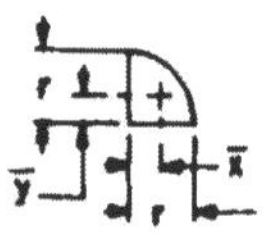
FILLET	$0.215r^2$	$0.774r$	$0.774r$	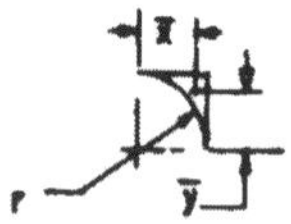

FIG. 6.7 Centroids of simple areas.

at the intersection of the X and Y axes. The horizontal and vertical distances to the centroid are indicated by $\bar{x}$ (read as x bar) and $\bar{y}$ (read as y bar) (Bassin et al., 1969).

We now explain the procedure for determining the centroids of composite areas (areas made up of combinations of the simple areas). The location of the centroid of a plane figure can be thought of as the average distance of the area to an axis. It is suggested that the tabular method shown be used in the following example problem. The composite figure is divided into numbered simple areas: rectangles, triangles, etc. The X and Y distances of the centroids of each of these simple areas from the intersection of the X and Y axes are recorded in the table.

EXAMPLE 6. Calculate $\bar{x}$ and $\bar{y}$ for the composite area shown in Fig. 6.8. The memories in which data will be stored are numbered in appropriate columns of the following table. All the entries in a column are summed and recalled as shown in the equations.

The symbol Δ denotes triangle. The symbol ϕ denotes diameter of a circle.

No.	Dimensions, in	Area, in² MeM A & D	x, in	Ax MeM B	y, in	Ay MeM C
1	3 × 3	9.0	1.5	13.5	1.5	13.5
2	Δ3 × 3	4.5	4.0	18.0	1.0	4.5
	ΣA =	13.5	ΣAx =	31.5	ΣAy =	18.0

$$\bar{x} = \frac{\Sigma Ax}{\Sigma A} = \frac{31.5}{13.5} = 2.333 \text{ in}$$

$$\bar{y} = \frac{\Sigma Ay}{\Sigma A} = \frac{18.0}{13.5} = 1.333 \text{ in}$$

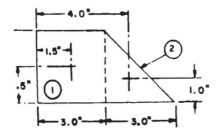

FIG. 6.8

Calculator routine for algebraic notation:

PRESS	DISPLAY	COMMENT
3 × 3 =	9	A 1
STO ALPHA D =		
STO ALPHA A = × 1.5 =	13.5	A 1 X 1
STO ALPHA B =		
ALPHA D × 1.5 =	13.5	A 1 Y 1
STO ALPHA C =		
3 × 3 ÷ 2 =	4.5	A 2
STO + ALPHA A = STO ALPHA E =		
× 4 = STO + ALPHA B =	18	A 2 X 2
ALPHA E × 1 =	4.5	A 2 Y 2
STO + ALPHA C =		
ALPHA B ÷ ALPHA A =	2.333	$\bar{x}$
ALPHA C ÷ ALPHA A =	1.333	$\bar{y}$

Calculator routine for reverse Polish notation:

PRESS	DISPLAY	COMMENT
3 ENTER 3 × STO A STO D	9	A 1
1.5 × STO B	13.5	A 1 X 1
RCL D 1.5 × STO C	13.5	A 1 Y 1
3 ENTER 3 × 2 ÷		
STO + A STO E	4.5	A 2
4 × STO + B	18	A 2 X 2
RCL E 1 × STO + C	4.5	A 2 Y 2
RCL B RCL A +	2.333	$\bar{x}$
RCL C RCL A +	1.333	$\bar{y}$

If a hole of any shape exists in a plane figure, treat it as a negative area.

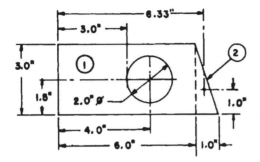

FIG. 6.9

EXAMPLE 7. Calculate $\bar{x}$ and $\bar{y}$ for the area shown in Fig. 6.9. Treat the 2-in diameter circle as a negative area.

No.	Dimensions, in	Area, in² MeM A	x, in	Ax MeM B	y, in	Ay MeM C
1	6 × 3	18.0	3.0	54.0	1.5	27.0
2	Δ1 × 3	1.5	6.33	9.5	1.0	1.5
3	2.0φ	$\underline{-3.14}$	4.0	$\underline{-12.57}$	1.5	$\underline{-4.71}$
		$\Sigma A = 16.36$		$\Sigma Ax = 50.93$		$\Sigma Ay = 23.79$

$$\bar{x} = \frac{\Sigma Ax}{\Sigma A} = \frac{50.93}{16.36} = 3.11 \text{ in}$$

$$\bar{y} = \frac{\Sigma Ay}{\Sigma A} = \frac{23.79}{16.36} = 1.45 \text{ in}$$

IV. Theorem of Pappus and Guldinus

The theorem of Pappus and Guldinus (Oberg and Jones, 1964) states that the volume of a solid body generated by the revolution of a plane surface about an axis is equal to the area of the surface multiplied by the length of the path of its center of gravity. If a complete revolution of the surface is made,

$$V = 2\pi RA \tag{1}$$

where

    R = radius to the centroid of the surface
    A = area of the surface
    V = volume generated

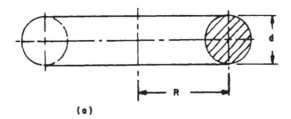

(a)

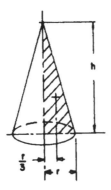

(b)

FIG. 6.10 (a) Volume of torus. (b) Volume of right circular cone.

According to this principle, the volume of a torus (see Fig. 6.10a) is equal to

$$2\pi R \frac{\pi}{4d^2} = \frac{\pi^2 R d^2}{2}$$

The volume of a right circular cone (see Fig. 6.10b) is equal to

$$2\pi \frac{r}{3} \frac{rh}{2} = \frac{\pi r^2 h}{3}$$

Using the table for centroids, it is possible to determine the volume of almost any area of revolution.

EXAMPLE 8. Calculate the volume in Fig. 6.11 generated by revolving the area ABCD around the axis Y-Y. Tabulate the procedure as shown below.

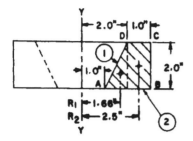

FIG. 6.11

Calculator routine for algebraic notation:

PRESS	DISPLAY	COMMENT
1 + (2 + 3) =	1.667	R 1
STO ALPHA A =	.	
2 + (1 + 2) =	2.5	R 2
STO ALPHA B =		
2 × 2nd π × ALPHA A =	10.47	V 1
STO ALPHA C =		
2 × 2 × 2nd π × ALPHA B =	31.42	V 2
STO + ALPHA C =		
RCL ALPHA C =	41.89	Σ V

Calculator routine for reverse Polish notation:

PRESS	DISPLAY	COMMENT
2 ENTER 3 + 1 × 1 +	1.667	R 1
STO A		
1 ENTER 2 + 1 × 2 +	2.5	R 2
STO B		
1 ENTER 2 × 2 +	1.0	A 1
2 × ⌐ π × RCL A × STO C	10.457	V 1
1 ENTER 2 ×	2.0	A 2
2 × ⌐ π × RCL B × STO + C	31.42	V 2
RCL C	41.89	Σ V

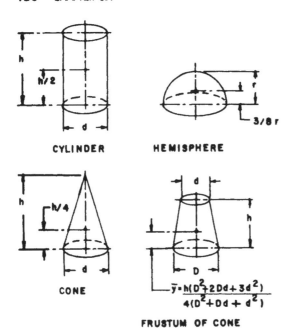

CYLINDER   HEMISPHERE

CONE

$$\bar{y} = \frac{h(D^2 + 2Dd + 3d^2)}{4(D^2 + Dd + d^2)}$$

FRUSTUM OF CONE

FIG. 6.12   Centers of gravity for simple volumes.

V. Center of Gravity

In calculating the center of gravity, the procedure is the same as for calculating centroids except that we are dealing with weights instead of areas. If the object for which the center of gravity is being calculated is entirely of one material, we can use volumes for the calculation. The centers of gravity for some simple volumes are given in Fig. 6.12.

EXAMPLE 9.   Calculate $\bar{y}$ for the homogeneous solid shown in Fig. 6.13. Fill in the tabulation as indicated and make use of the calculator's memory capacity.

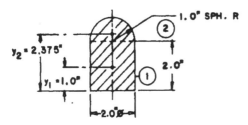

FIG. 6.13

No.	Dimensions	MeM A, V	y	MeM B, Vy
1	Cylinder, 2-in diameter × 2 in	6.283	1.0	6.283
2	Hemisphere, 1-in radius	2.095	2.37	4.976
		ΣV = 8.378		ΣVy =11.259

$$\bar{y} = \frac{\Sigma Vy}{\Sigma V} = \frac{11.259}{8.378} = 1.344 \text{ in}$$

Calculator routine for algebraic notation:

PRESS	DISPLAY	COMMENT
2 + (3 ÷ 8 × 1) =	2.375	Y 2
STO ALPHA A =		
2 × 2nd π × 2 x² ÷ 4 =	6.283	V 1
STO ALPHA B = STO ALPHA C =		
2 ÷ 3 × 2nd π × (1 yˣ 3) =	2.094	V 2
STO + ALPHA B = STO ALPHA D =		
1 × ALPHA C =	6.283	V 1 Y 1
STO ALPHA E =		
ALPHA D × ALPHA A =	4.974	V 2 Y 2
STO + ALPHA E =		
ALPHA E ÷ ALPHA B =	1.343	$\bar{y}$

Calculator routine for reverse Polish notation:

PRESS	DISPLAY	COMMENT
3 ENTER 8 ÷ 1 × 2 +	2.375	Y 2
STO A		
2 ENTER ⌜ π × 2 ⌝ x² × 4 ÷	6.283	V 1
STO B STO D		

PRESS	DISPLAY	COMMENT
3 <u>ENTER</u> 3 + Γ π X		
1 <u>ENTER</u> 3 y^x X	2.094	V 2
<u>STO</u> + B <u>STO</u> E		
<u>RCL</u> D 1 X <u>STO</u> C	6.283	V 1 Y 1
<u>RCL</u> E <u>RCL</u> A X <u>STO</u> + C	4.974	V 2 Y 2
<u>RCL</u> C <u>RCL</u> B +	1.343	$\bar{y}$

If the object for which the center of gravity is being calculated is composed of two or more different materials, the weight of each material must be calculated from the product of the volume and the density of the material.

EXAMPLE 10. Calculate $\bar{y}$ for the solid figure composed of two different materials shown in Fig. 6.14. Fill in the tabulation as indicated and make use of the calculator's memory capacity.

No.	Dimensions	V	D	W	y	Wy
1	Cone, 2-in diameter × 4 in	4.188	0.283	1.185	3.0	3.555
2	Hemisphere, 0.75-in radius	−0.884	0.283	−0.25	3.719	−0.93
3	Hemisphere, 0.75-in radius	0.884	0.412	<u>0.364</u>	3.719	<u>1.354</u>
				$\Sigma W = 1.299$		$\Sigma Wy = 3.979$

$$\bar{y} = \frac{\Sigma Wy}{\Sigma W} = \frac{3.979}{1.299} = 3.063 \text{ in.}$$

The examples in this chapter have illustrated the basic procedure for their solution. Many other kinds of figures were not covered. Information on these can be obtained in any good engineering handbook. The exercises in this chapter should help develop skill in making use of the calculator's memory capabilities. This skill will save time and will give more accurate answers to problems. It will also be helpful later when the subject of programming is taken up.

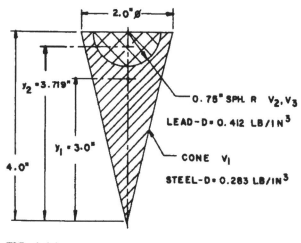

FIG. 6.14

Exercise 6.1

6.1-1 A spherical gal tank has a volume of 100,000 gal.

(a) Calculate the diameter, in feet.

(b) Calculate the surface area, in square feet.

6.1-2 The forging shown in Fig. 6.15 is to be formed from a length of a 4-in-diameter bar called a slug.

(a) Calculate the volume, in cubic inches.

(b) Calculate the length of the slug, in inches.

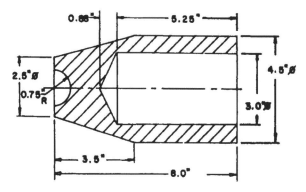

FIG. 6.15

6.1-3 A 30-in-diameter drum 50 in long lies in a horizontal position. Calculate the volume of fluid at depths of 5 in, 10 in, 15 in, 20 in, 25 in.

6.1-4 In Fig. 6.16 are shown two sections 100 ft apart of a site to be excavated. Calculate the volume of earth (in cubic yards) which must be removed.

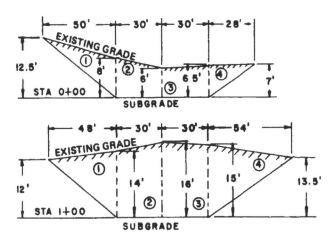

FIG. 6.16

Exercise 6.2

Calculate the centroids of the following figures:

6.2-1 The angle shown in Fig. 6.17.

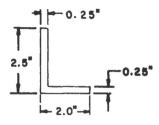

FIG. 6.17

6.2-2 The tee shown in Fig. 6.18.

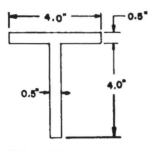

FIG. 6.18

6.2-3 The hat section shown in Fig. 6.19.

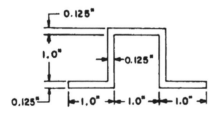

FIG. 6.19

6.2-4 The area shown in Fig. 6.20.

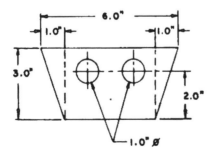

FIG. 6.20

6.2-5 The channel shown in Fig. 6.21.

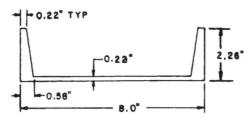

FIG. 6.21

6.2-6 The area shown in Fig. 6.22.

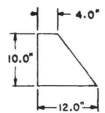

FIG. 6.22

6.2-7 The area shown in Fig. 6.23.

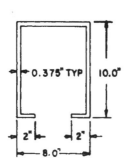

FIG. 6.23

6.2-8 Derive the formula for the centroid of a fillet of radius r, using a suare with dimensions r × r enclosing a negative area of a quadrant of radius r.

Exercise 6.3

Using the theorem of Pappus and Guldinus, determine the volume of the following:

6.3-1   The figure shown in Fig. 6.24.

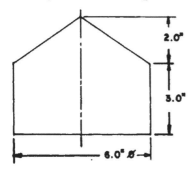

FIG. 6.24

6.3-2   A cylinder of radius r and height h.
6.3-3   A sphere of radius r.
6.3-4   The figure shown in Fig. 6.25.

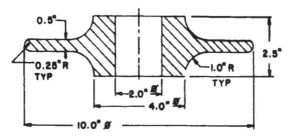

FIG. 6.25

6.3-5   The figure shown in Fig. 6.26.

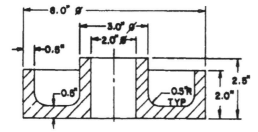

FIG. 6.26

Exercise 6.4

Calculate the center of gravity $\bar{y}$ of the following:

6.4-1   The figure shown in Fig. 6.27.

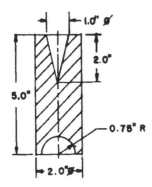

FIG. 6.27

6.4-2   The figure shown in Fig. 6.28.

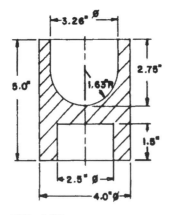

FIG. 6.28

6.4-3   The figure shown in Fig. 6.29.

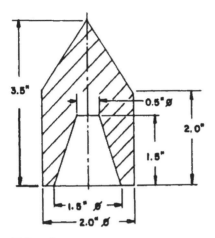

FIG. 6.29

6.4-4   The figure shown in Fig. 6.30.

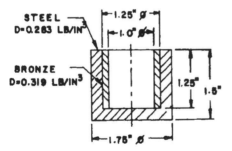

FIG. 6.30

7 · Logarithms and Exponentials

I. Common Logarithms

A logarithm is a power to which a certain base number must be raised to equal a given number. There are two bases in general use. Logarithms to base 10 are called common logarithms and are determined on your calculator by pressing the "LOG" key. On the Hewlett Packard HP-32SII it is necessary to press "◄⌐" before pressing the "LOG" key (Hewlett Packard, 1990). Logarithms to base e are called natural logarithms and are determined on your calculator by pressing the "LN" key. Natural logarithms will be explained more fully in Section III.

For many years logarithms were used for complex calculations, by means of tables of logarithms, before calculators and computers were available. The general laws of logarithms are (Washington, 1970):

$$\log A \times B = \log A + \log B$$

$$\log \frac{A}{B} = \log A - \log B$$

$$\log C^N = N \log C$$

$$\log \sqrt[N]{C} = \frac{\log C}{N}$$

Although it is no longer necessary to use logarithms to perform these operations, knowledge of the function of logarithms is still needed to solve many mathematical problems. We will work through some basic mathematical calculations with the calculator, using logarithms.

EXAMPLE 1. $250 \times 30 = 7500$.

Calculator routine for algebraic notation:

PRESS	DISPLAY	COMMENT
2nd 10ˣ		antilog key
(LOG 250 + LOG 30) =	7500	product

Calculator routine for reverse Polish notation:

PRESS	DISPLAY	COMMENT
250 ↰ LOG		
30 ↰ LOG +	3.8751	log product
↰ 10ˣ	7500.0	product

Note that the answer is called the antilog of the log product. Note also that the part of the logarithm to the left of the decimal point is the power of 10 of the number. It is called the *characteristic*. The part of the logarithm to the right of the decimal point is called the *mantissa* and is determined by the actual digits of the number. The mantissa is the same for 3, 30, 300, etc. Now let us explain the relationship for numbers less than 10. You will remember that the powers of 10 are negative for numbers less than 1. It is impossible for the calculator to show a negative characteristic with a positive mantissa. For example, the logarithm of 2 is 0.30103. For a number less than 1 the characteristic is subtracted from the mantissa. Therefore the logarithm of 0.2 is equal to $0.30103 - 1 = -0.69897$.

EXAMPLE 2. $0.25 \times 0.003 = 0.00075$.

Calculator routine for algebraic notation:

PRESS	DISPLAY	COMMENT
2nd 10^x		antilog key
(LOG 0.25 + LOG 0.003) =	0.00075	product

Calculator routine for reverse Polish notation:

PRESS	DISPLAY	COMMENT
0.25 ⌐ LOG		
0.003 ⌐ LOG +	-3.1349	log product
⌐ 10^x	0.00075	product

EXAMPLE 3. $375 \div 0.625 = 600$.

Calculator routine for algebraic notation:

PRESS	DISPLAY	COMMENT
2nd 10^x		antilog key
(LOG 375 − LOG 0.625) =	600.0	quotient

Calculator routine for reverse Polish notation:

PRESS	DISPLAY	COMMENT
375 ⌐ LOG		
0.625 ⌐ LOG −	2.7782	log quotient
10^x⌐	600.0	quotient

EXAMPLE 4. $\sqrt[3]{935.6} = 9.7805529$.

Calculator routine for algebraic notation:

PRESS	DISPLAY	COMMENT
2nd 10^x		antilog key
(LOG 935.6 + 3) =	9.78055	antilog

Calculator routine for reverse Polish notation:

PRESS	DISPLAY	COMMENT
935.6 ↰ <u>LOG</u>		
3 ÷ ↰ 10^x	9.78055	antilog

II. Applications of Common Logarithms

Most of the important applications of common logarithms occur in the evaluation of data varying over an extremely wide range or data involving very large numbers. A good example is the decibel (dB), used in measuring the intensity of sound. The value of the decibel is related to the intensity of power at the threshold of hearing, which is 10^{-10} μW/cm². For each time that the power delivered to the ear is multiplied by 10, the sound intensity is increased by 10 dB. When the sound intensity increases from 0 dB at the threshold of hearing to 120 dB, the power reaching the ear is increased 10^{12} times! Loudness in decibels is defined as (Harris and Hemmerling, 1972):

$$L = 10 \log \frac{I_1}{I_2} \tag{1}$$

where

L = loudness, dB

I_1 = intensity of sound at one distance

I_2 = intensity of sound at a second distance

The formula which relates the intensity of sound and distance from the source is

$$\frac{I_1}{I_2} = \frac{s_2^2}{s_1^2} \tag{2}$$

where

s_1 = the first distance

s_2 = the second distance

EXAMPLE 5. The intensity of sound from a siren is 120 dB at a distance of 500 ft. What is the intensity at a distance of 1 mi?

Transposing equation (2) to solve for I_2:

$$I_2 = \frac{I_1 s_1^2}{s_2^2} = \frac{10^{12} \times 500^2}{5280^2} = 8.967E9$$

$10 \log 8.967E9 = 99.53$ dB

Because this calculation involves scientific notation, it will be necessary to show routines for all three calculators.

Calculator routine for the TI-68:

PRESS	DISPLAY	COMMENT
LOG (1 EE 12 × 5000 x^2		
÷ 5280 x^2) × 10 =	99.53	I_2, dB

Calculator routine for the FX-7000GA:

PRESS	DISPLAY	COMMENT
MODE 8 , 4 EXE		
LOG (1 EXP 12 × 500 x^2		
÷ 5280 x^2) × 10 EXE	99.53 E 01	I_2, dB

Calculator routine for the 32SII:

PRESS	DISPLAY	COMMENT
⌐ DISP (SC) 5		
1 E 12 ENTER		
500 ⌐ x^2 × 5280 ⌐ x^2 ÷		
⌐ LOG 10 ×	9.953 E 1	I_2, dB

The measurement of sound intensity has become very important in engineering. A great deal of research goes into the development of means to control the noise level in vehicles. New industrial safety standards require intensive study of the sound levels in industrial plants.

A similar use of logarithmic measurement is made in the Richter scale for rating the intensity of earthquakes. Following is a formula for comparison of the intensity of earthquakes (Hewlett-Packard, 1978a):

$$R_2 - R_1 = \log \frac{M_2}{M_1} \tag{3}$$

where

R_2 and R_1 = respective Richter scale readings

$\log \dfrac{M_2}{M_1}$ = log of the ratio of severity

EXAMPLE 6. The 1906 San Francisco earthquake had an intensity of 8.5 on the Richter scale. How much more severe was it than an earthquake registering 6.5 on the Richter scale? Transposing equation (3):

$$\log \frac{M_2}{M_1} = 8.5 - 6.5 = 2$$

antilog 2 = 100

Calculator routine for algebraic notation:

PRESS	DISPLAY	COMMENT
2nd 10^x (8.5 − 6.5) =	100	Ratio M_2/M_1

Calculator routine for reverse Polish notation:

PRESS	DISPLAY	COMMENT
⌐ DISP (FX) 2		
8.5 ENTER 6.5 −	2	LOG M_2/M_1
⌐ 10^x	100	Ratio M_2/M_1

Another important application of logarithms is in chemistry, where the acidity or alkalinity of a solution is designated by the term pH. The pH is the logarithm of the reciprocal of the hydrogen-ion concentration, when this concentration is expressed as moles per liter. This is equivalent to defining the pH as the negative of the logarithm of the hydrogen-ion concentration. A solution whose pH is 7 is neutral. A solution with pH greater than 7 is alkaline. One with a pH less than 7 is acidic (Sorum, 1969).

EXAMPLE 7. Calculate the pH of a solution which contains 2.5E-4 moles of H^+ per liter.

Calculator routine for algebraic notation:

PRESS	DISPLAY	COMMENT
LOG (2.5 EE ($-$)4 x^{-1}) =	3.6	pH

Calculator routine for reverse Polish notation:

PRESS	DISPLAY	COMMENT
⌐ DISP (SC) 2		
2.5 E +/ − 4 $\frac{1}{x}$		
⌐ LOG	3.6	pH

EXAMPLE 8. Calculate the H^+ ion concentration in moles per liter of a solution whose pH is 8.5.

Calculator routine for algebraic notation:

PRESS	DISPLAY	COMMENT
3rd ScEn)		
2nd 10^x 8.5 =	3.16 E 08	Antilog
x^{-1} =	3.16 E − 09	H+ conc.

Calculator routine for reverse Polish notation:

PRESS	DISPLAY	COMMENT
⌐ DISP (SC) 2		
8.5 ⌐ 10^x	3.16 E 8	Antilog
$\frac{1}{x}$	3.16 E −9	H+ conc.

III. Natural Logarithms

Many quantities in nature grow in much the same way as a sum of money at compound interest. While money grows step by step at the end of each interest conversion period, growth or decay in nature is usually a continuous process. The number of interest periods could be considered to be infinite.

The mathematical constant e, which is equal to 2.7182818 to eight figures, occurs frequently in equations concerned with natural growth or decay, such as the speed of a chemical reaction, the rate of cooling of a warm object, etc. Mathematically, e is derived from the expansion of the series

$$e = 1 + \frac{1}{1} + \frac{1}{2!} + \frac{1}{3!} + \frac{1}{4!} \cdots \frac{1}{n!} \tag{4}$$

The expression 3! is read as 3 factorial and is equal to the product of all the integers up to and including 3.

$$3! = 1 \times 2 \times 3$$

The calculation of the factorial of a number can be accomplished by pressing the x! key on your calculator.

Equations involving e have the general form: $x = e^n$. Because n in this equation usually involves some fractional number such as 0.35 or 1.4, the solution of this equation has traditionally involved the use of logarithms, since they provided the simplest method of raising a number to a fractional power. Consequently, equations involving natural growth and decay use the logarithm to the base e. This is abbreviated ln and is called the natural logarithm.

EXAMPLE 9. ln 75.45 = 4.3234702.

Calculator routine for the TI-68:

PRESS	DISPLAY
2nd FIX 5	
LN 75.45 =	4.32347

Calculator routine for the FX-7000GA:

PRESS	DISPLAY
MODE 7 , 5	
LN 75.45 EXE	4.32347

Calculator routine for the HP-32SII:

PRESS	DISPLAY
⌐ DISP (FX) 5	
75.45 LN	4.32347

EXAMPLE 10. $x = e^{0.3} = 1.3498588$.

Calculator routine for algebraic notation:

PRESS	DISPLAY
2nd e^x 0.3 =	1.34986

Calculator routine for reverse Polish notation:

PRESS	DISPLAY
⌐ DISP (FX) 5	
0.3 e^x	1.34986

In equations of decay, the general formula becomes $x = e^{-n}$.

EXAMPLE 11. $x = e^{-0.4} = 0.67032$.

Calculator routine for algebraic notation:

PRESS	DISPLAY
2nd e (−) 0.4 =	0.67032

Calculator routine for reverse Polish notation:

PRESS	DISPLAY
⌐ DISP (FX) 5	
0.4 +/− e^x	0.67032

IV. Applications of Natural Logarithms

A common application of natural logarithms is in determining the relative tension between the tight side and the slack side of a rope or belt wrapped around a drum (see Fig. 7.1).

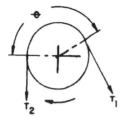

FIG. 7.1 Belt tension around a drum.

The formula is

$$\frac{T_1}{T_2} = e^{f\Theta} \qquad (5)$$

where

T_1 = tight side tension
T_2 = slack side tension
f = coefficient of friction
Θ = wrap around the drum, rad

This formula is used in determining the relative tension of the two sides of a belt drive or of a band brake.

EXAMPLE 12. In a belt drive the wrap of the belt around a pulley is 150° and the coefficient of friction is 0.3. Determine the tension ratio T_1/T_2.

Calculator routine for algebraic notation:

PRESS	DISPLAY
2nd e^x (150 × 2nd π +	
180 × 0.3) =	2.19

Calculator routine for reverse Polish notation:

PRESS	DISPLAY
⌐ DISP (FX) 2	
150 Γ →RAD	
0.3 × e^x	2.19

Now let us consider a similar problem from another angle. Suppose that the tension ratio and coefficient of friction are known and that it is desired to find the minimum allowable wrap.

If $T_1/T_2 = e^{f\Theta}$, then $f\Theta = \ln T_1/T_2$ and $\Theta = \ln (T_1/T_2)/f$

EXAMPLE 13. Find the minimum permissible wrap in degrees for a belt drive if $T_1/T_2 = 2$ and $f = 0.3$.

Calculator routine for algebraic notation:

PRESS	DISPLAY	COMMENT
<u>LN</u> 2 + 0.3 =	2.31	Wrap, rad.
× 180 ÷ <u>2nd</u> π =	132.4	Wrap, deg.

Calculator routine for reverse Polish notation:

PRESS	DISPLAY	COMMENT
⌐ <u>DISP</u> (<u>FX</u>) 2		
2 <u>LN</u> <u>ENTER</u> 0.3 ÷	2.31	Wrap, rad.
⌐→ <u>DEG</u>	132.38	Wrap, deg.

An approximation for determining altitude is given by the following formula (Hewlett-Packard, 1978a):

$$A = 25,000 \ln(P_1/P_2) \tag{6}$$

where

A = altitude, ft
P_1 = barometer reading at sea level, in
P_2 = barometer reading at location, in

EXAMPLE 14. The barometer reading at the top of a mountain is 20.0 in, while that at sea level is 29.92 in. How high is the mountain?

Calculator routine for algebraic notation:

PRESS	DISPLAY	COMMENT
<u>2nd</u> <u>FIX</u> 2		
<u>LN</u> (29.92 ÷ 20) =	1.5	P_1/P_2
× 25000 =	10069.9	Altitude, ft.

Calculator routine for reverse Polish notation:

PRESS	DISPLAY	COMMENT
⤺ DISP (FX) 2		
29.92 ENTER 20 ÷		
LN 25000 ×	10069.9	Altitude, ft.

Now let us consider some equations of decay. Remember that the general form of these equations is $x = e^{-n}$. This could also be written as $x = 1/e^n$. Following is the equation for determining the speed of a free-spinning wheel after time t:

$$V_2 = V_1 e^{-rt} \tag{7}$$

where

V_2 = final rpm
V_1 = initial rpm
r = % decrease per minute
t = time, min

EXAMPLE 15. A free-spinning wheel is turning at 1500 rpm. What is its rpm after 10 min if r is 35%/min?

$$V_2 = V_1 e^{-rt} = 1500e^{-(0.35)(10)} = 45.3 \text{ rpm}$$

Calculator routine for algebraic notation:

PRESS	DISPLAY	COMMENT
2nd e^x (−) (0.35 × 10)		
× 1500 =	45.29	V 2, RPM

Calculator routine for reverse Polish notation:

PRESS	DISPLAY	COMMENT
⤺ DISP (FX) 2		
0.35 ENTER 10 × +/−		
e^x 1500 ×	45.29	V 2, RPM

Now suppose we turn the problem around and solve for the time required to reach a certain speed.

If $V_2 = V_1 e^{-rt} = V_1/e^{rt}$, then $V_1/V_2 = e^{rt}$ and $\ln(V_1/V_2) = rt$.

EXAMPLE 16. A free-spinning wheel is turning at 1,500 rpm. How many minutes will it take to slow to 500 rpm if r is 35%/min?

$$t = \frac{\ln(V_1/V_2)}{r} = \frac{\ln(1500/500)}{0.35} = 3.139 \text{ min}$$

Calculator routine for algebraic notation:

PRESS	DISPLAY	COMMENT
LN (1500 + 500)		
+ 0.35 =	3.14	Time, min.

Calculator routine for reverse Polish notation:

PRESS	DISPLAY	COMMENT
⅂ DISP (FX) 2		
1500 ENTER 500 + LN		
0.35 +	3.14	Time, min.

Radium decomposes according to the following equation:

$$A = Me^{-rt} \qquad (8)$$

where

A = remaining mass, mg
M = original mass, mg
r = rate per century, 4.1%
t = time, centuries

EXAMPLE 17. If the original mass of radium is 25 mg, how much will remain after 2000 years?

Applying equation (8),

$$A = 25e^{-(0.041)(20)} = 11.01 \text{ mg}$$

Calculator routine for algebraic notation:

PRESS	DISPLAY	COMMENT
20 × 2nd e^x (–) (0.041		
× 20) =	11.01	M, mg

Calculator routine for reverse Polish notation:

PRESS	DISPLAY	COMMENT
⌐ DISP (FX) 2		
0.041 ENTER 20 ×		
+ / – e^x 25 ×	11.01	M, mg

The examples shown in this chapter are intended to give a basic idea of the scope and application of problems involving logarithms. A much broader understanding of this subject will be obtained in more advanced courses.

Exercise 7.1

Solve the following problems, using common logarithms:

7.1-1 $375 \times 0.056 =$

7.1-2 $9756 \div 893 =$

7.1-3 $\dfrac{953.5 \times 896.2}{3525} =$

7.1-4 $\sqrt{9756.4} =$

7.1-5 $\sqrt[3]{875 \times 32.55} =$

7.1-6 $\sqrt{(25.75)^2 + (39.63)^2} =$

7.1-7 $1.563 \times \sin 25.5° =$

7.1-8 $\dfrac{x}{\sin 35°} = \dfrac{355}{\sin 45°}$

7.1-9 $\pi/4 \times (0.375)^2 =$

7.1-10 $(25)^2 + (15)^3 + \sqrt{575} =$

Exercise 7.2

7.2-1 The intensity of sound from a milling machine is 100 dB at a distance of 3 ft. How far from the machine would one have to stand to feel an intensity of 80 dB?

7.2-2 The intensity of sound from a jet plane is 120 dB at 100 ft. What is its intensity at 2000 ft?

7.2-3 A severe earthquake registers 8.0 on the Richter scale. How much more severe was it than one registering 4.5 on the Richter scale?

7.2-4 The intensity of sound at 1000 ft from the source is 92 dB. What is its intensity at 50 ft from the source?

7.2-5 Calculate the pH of a solution which has 3.25E-4 moles of H_+ per liter.

7.2-6 Calculate the H_+ ion concentration in moles per liter of a solution whose pH is 6.

Exercise 7.3

7.3-1 Using the factorial function, derive the value of e by expanding the series:

$$e = 1 + \frac{1}{1!} + \frac{1}{2!} + \frac{1}{3!} + \cdots + \frac{1}{10!}$$

7.3-2 In a belt drive the belt wraps around the pulley 160°. (See Fig. 7.2.) If the coefficient of friction is 0.25, what is the ratio T_1/T_2?

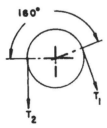

FIG. 7.2

7.3-3 How many turns around a drum must a rope be wrapped for $T_1/T_2 = 10$ if f = 0.2?

7.3-4   If the ratio $T_1/T_2 = 3$ and the wrap of a belt around a pulley is $180°$, what coefficient of friction is required?

7.3-5   The barometer reading at the top of a mountain is 15 in of mercury while that at sea level is 29.92 in. How high is the mountain?

7.3-6   If the barometer reading at sea level is 29.92 in of mercury, what should it be at 20,000 ft?

7.3-7   A free-spinning wheel is turning at 10,000 rpm. What is its rpm after 15 min if $r = 20\%/min$?

7.3-8   If a free-spinning wheel is turning at 5000 rpm, how many minutes will it take to slow to 50 rpm if $r = 20\%/min$?

7.3-9   A mass of 10 mg of radium will decay at the rate of 4.1% per century. How much will remain after 500 years?

8 · Applications of Higher Mathematics

I. Determinants

Many mathematical problems involve solving for two or more unknowns. If simultaneous equations can be written involving the unknown quantities, a number of equations equal to the number of unknowns will yield a solution. There are a number of methods for solving simultaneous equations, but the method using determinants yields the simplest solution when a calculator is available.

A. Second-Order Determinants

Setting up the coefficients of the unknowns in a pair of simultaneous equations in an array according to a simple set of rules is the basis of determinants. Consider the following two equations (Washington, 1970):

$$a_1x + b_1y = c_1$$

$$a_2x + b_2y = c_2$$

Arranging the coefficients of x and y, the denominator of the determinant becomes

which can be evaluated as $a_1b_2 - a_2b_1$ where the product of the coefficients in the upward-pointing diagonal is subtracted from that of the coefficients in the downward-pointing diagonal.

Solutions for x and y are as follows:

$$x = \frac{\begin{vmatrix} c_1 & b_1 \\ c_2 & b_2 \end{vmatrix}}{\begin{vmatrix} a_1 & b_1 \\ a_2 & b_2 \end{vmatrix}} = \frac{c_1b_2 - c_2b_1}{a_1b_2 - a_2b_1}$$

$$y = \frac{\begin{vmatrix} a_1 & c_1 \\ a_2 & c_2 \end{vmatrix}}{\begin{vmatrix} a_1 & b_1 \\ a_2 & b_2 \end{vmatrix}} = \frac{a_1c_2 - a_2c_1}{a_1b_2 - a_2b_1}$$

We can see from these solutions that the determinant of the denominator is made up of the coefficients of x and y. The determinant of the numerator for the solution of x is obtained by placing the column of c's on the left and the column of b's on the right. The determinant of the numerator for the solution of y is obtained by placing the column of a's on the left and the column of c's on the right.

EXAMPLE 1. Solve the following system of equations, using determinants, and check by substituting the solved values of x and y in one of the original equations:

$$2x + y = 3$$

$$5x + 3y = 10$$

$$x = \frac{\begin{vmatrix} 3 & 1 \\ 10 & 3 \end{vmatrix}}{\begin{vmatrix} 2 & 1 \\ 5 & 3 \end{vmatrix}} = \frac{9 - 10}{6 - 5} = \frac{-1}{1} = -1$$

$$y = \frac{\begin{vmatrix} 2 & 3 \\ 5 & 10 \end{vmatrix}}{\begin{vmatrix} \text{Denom-} \\ \text{inator} \end{vmatrix}} = \frac{20 - 15}{1} = 5$$

Calculator routine for algebraic notation:

PRESS	DISPLAY	COMMENT
$3 \times 3 - 10$		
$\times 1 =$ <u>STO</u> <u>ALPHA</u> A $=$	-1	Numerator, X
$2 \times 3 - 5$		
$\times 1 =$ <u>STO</u> <u>ALPHA</u> B $=$	1	Denominator
<u>ALPHA</u> A $\div$ <u>ALPHA</u> B $=$		
<u>STO</u> <u>ALPHA</u> C $=$	-1	X
$2 \times 10 - 5 \times 3$		
$\div$ <u>ALPHA</u> B $=$ <u>STO</u> <u>ALPHA</u> C $=$	5	Y
$2 \times$ <u>ALPHA</u> A $+$ <u>ALPHA</u> C $=$	3	Check

Calculator routine for reverse Polish notation:

PRESS	DISPLAY	COMMENT
⌐ <u>DISP</u> (<u>FX</u>) 2		
3 <u>ENTER</u> 3 $\times$ 10 <u>ENTER</u>		
1 x $-$	-1	Numerator, X
2 <u>ENTER</u> 3 $\times$ 5 <u>ENTER</u>		
1 $\times$ $-$ <u>STO</u> A	1	Denominator
$\div$ <u>STO</u> B	-1	X
2 <u>ENTER</u> 10 $\times$ 5 <u>ENTER</u>		
3 x $-$	5	Numerator, Y
<u>RCL</u> A $\div$	5	Y
<u>RCL</u> B 2 x $+$	3	Check

EXAMPLE 2. Solve the following system of equations and check by substituting solved values of x and y in one of the original equations:

$$7x + y = 3$$
$$5x - 2y = 0$$

$$x = \frac{\begin{vmatrix} 3 & 1 \\ 0 & -2 \\ 7 & 1 \\ 5 & -2 \end{vmatrix}}{} = \frac{-6}{-19} = 0.3157895 \text{ (round off to } 0.3158)$$

$$y = \frac{\begin{vmatrix} 7 & 3 \\ 5 & 0 \end{vmatrix}}{\begin{matrix} \text{Denom-} \\ \text{inator} \end{matrix}} = \frac{-15}{-19} = 0.7894737 \text{ (round off to } 0.7895)$$

Note that in this case it is necessary to retain the full values of x and y in memory to check. The rounded-off values will not give a perfect check. Also note that the product of a diagonal containing a zero is equal to zero. The calculator routine is the same as that for the previous example.

B. Third-Order Determinants

For solving three equations with three unknowns the procedure for arranging determinants is similar to that for second-order determinants. Consider the following three equations:

$$a_1x + b_1y + c_1z = d_1$$

$$a_2x + b_2y + c_2z = d_2$$

$$a_3x + b_3y + c_3z = d_3$$

The determinant for the denominator is arranged as follows:

$$= a_1b_2c_3 + b_1c_2a_3 + c_1a_2b_3$$
$$- a_3b_2c_1 - b_3c_2a_1 - c_3a_2b_1$$

Note that columns 1 and 2 are repeated and that the value of the determinant is equal to the sum of the products in each of the downward-pointing diagonals minus the sum of the products in each of the upward-pointing diagonals.

$$x = \frac{d_1b_2c_3 + b_1c_2d_3 + c_1d_2b_3 - d_3b_2c_1 - b_3c_2d_1 - c_3d_2b_1}{a_1b_2c_3 + b_1c_2a_3 + c_1a_2b_3 - a_3b_2c_1 - b_3c_2a_1 - c_3a_2b_1}$$

$$y = \frac{\text{Denominator}} = \frac{a_1d_2c_3 + d_1c_2a_3 + c_1a_2d_3 - a_3d_2c_1 - d_3c_2a_1 - c_3a_2d_1}{\text{Denominator}}$$

$$z = \frac{\text{Denominator}} = \frac{a_1b_2d_3 + b_1d_2a_3 + d_1a_2b_3 - a_3b_2d_1 - b_3d_2a_1 - d_3a_2b_1}{\text{Denominator}}$$

EXAMPLE 3. Solve the following system of equations and check by substituting the solved values in one of the original equations:

$$2x + 3y + z = 7$$

$$3x - y - 2z = 17$$

$$4x + 5y + 3z = 7$$

$$x = \frac{\begin{matrix} d & b & c \\ 7 & 3 & 1 \\ 17 & -1 & -2 \\ 7 & 5 & 3 \end{matrix} \begin{matrix} 7 & 3 \\ 17 & -1 \\ 7 & 5 \end{matrix}}{\begin{matrix} a & b & c \\ 2 & 3 & 1 \\ 3 & -1 & -2 \\ 4 & 5 & 3 \end{matrix} \begin{matrix} 2 & 3 \\ 3 & -1 \\ 4 & 5 \end{matrix}} = \frac{-54}{-18} = 3$$

$$y = \frac{\begin{matrix} a & d & c \\ 2 & 7 & 1 \\ 3 & 17 & -2 \\ 4 & 7 & 3 \end{matrix} \begin{matrix} 2 & 7 \\ 3 & 17 \\ 4 & 7 \end{matrix}}{\text{Denominator}} = \frac{-36}{-18} = 2$$

$$z = \frac{\begin{matrix} a & b & d \\ 2 & 3 & 7 \\ 3 & -1 & 17 \\ 4 & 5 & 7 \end{matrix} \begin{matrix} 2 & 3 \\ 3 & -1 \\ 4 & 5 \end{matrix}}{\text{Denominator}} = \frac{90}{-18} = -5$$

EXAMPLE 4. Solve the following system of equations and check by substituting the solved values in one of the original equations:

$$2x + 3y + z = 10$$
$$x + 2y + z = 5$$
$$3x - 5y + z = 0$$

As the answers in this example are fractional, substitution in the third equation will not yield exactly zero. The calculator will give the answer 1E-11, which is a very small variation indeed.

$$x = \cfrac{\begin{vmatrix} d & b & c \\ 10 & 3 & 1 \\ 5 & 2 & -1 \\ 0 & -5 & 1 \end{vmatrix}\begin{matrix} 10 & 3 \\ 5 & 2 \\ 0 & -5 \end{matrix}}{\begin{vmatrix} a & b & c \\ 2 & 3 & 1 \\ 1 & 2 & -1 \\ 3 & -5 & 1 \end{vmatrix}\begin{matrix} 2 & 3 \\ 1 & 2 \\ 3 & -5 \end{matrix}} = \frac{-70}{-29}$$

$$= 2.4137931 \text{ (round off to 2.414)}$$

$$y = \cfrac{\begin{vmatrix} a & d & c \\ 2 & 10 & 1 \\ 1 & 5 & -1 \\ 3 & 0 & 1 \end{vmatrix}\begin{matrix} 2 & 10 \\ 1 & 5 \\ 3 & 0 \end{matrix}}{\text{Denominator}} = \frac{-45}{-29} = 1.5517241 \text{ (round off to 1.552)}$$

$$z = \cfrac{\begin{vmatrix} a & b & d \\ 2 & 3 & 10 \\ 1 & 2 & 5 \\ 3 & -5 & 0 \end{vmatrix}\begin{matrix} 2 & 3 \\ 1 & 2 \\ 3 & -5 \end{matrix}}{\text{Denominator}} = \frac{-15}{-29}$$

$$= 0.5172414 \text{ (round off to 0.5172)}$$

It is not essential that all three unknowns appear in each of the three equations. If one of the unknowns does not appear in an equation, enter the coefficient of that unknown as a zero in the determinant. If the value of the determinant of the denominator is zero and that of the numerator is not zero, the system is inconsistent. If the value of the numerator only is zero, the variable evaluates as zero.

EXAMPLE 5. Solve the following system of equations, and check by substituting the sovled values in one of the original equations:

$$x + y + z = 0$$
$$2x - y = 5$$
$$ 2y + 4z = 2$$

$$x = \frac{\begin{vmatrix} d & b & c \\ 0 & 1 & 1 \\ 5 & -1 & 0 \\ 2 & 2 & 4 \end{vmatrix} \begin{matrix} 0 & 1 \\ 5 & -1 \\ 2 & 2 \end{matrix}}{\begin{vmatrix} a & b & c \\ 1 & 1 & 1 \\ 2 & -1 & 0 \\ 0 & 2 & 4 \end{vmatrix} \begin{matrix} 1 & 1 \\ 2 & -1 \\ 0 & 2 \end{matrix}} = \frac{-8}{-8} = 1$$

$$y = \frac{\begin{vmatrix} a & d & c \\ 1 & 0 & 1 \\ 2 & 5 & 0 \\ 0 & 2 & 4 \end{vmatrix} \begin{matrix} 1 & 0 \\ 2 & 5 \\ 0 & 2 \end{matrix}}{\text{Denominator}} = \frac{24}{-8} = -3$$

$$z = \frac{\begin{vmatrix} a & b & d \\ 1 & 1 & 0 \\ 2 & -1 & 5 \\ 0 & 2 & 2 \end{vmatrix} \begin{matrix} 1 & 1 \\ 2 & -1 \\ 0 & 2 \end{matrix}}{\text{Denominator}} = \frac{-16}{-8} = 2$$

II. Applications of Determinants

One of the most common applications of determinants is in the solution of problems involving mixtures.

EXAMPLE 6. By weight one alloy is composed of 80% copper and 20% zinc. Another alloy is composed of 30% copper and 70% zinc. How many grams of each of these alloys would be required to make 500 g of an alloy which is 60% copper and 40% zinc?

Solution: Let

x = number of grams of first alloy required
y = number of grams of second alloy required

Then $x + y = 500$. The final alloy contains $0.6 \times 500 = 300$ g of copper. The weight of copper in the first alloy is $0.8x$ and the weight of copper in the second alloy is $0.3y$. Then $0.8x + 0.3y = 300$.

The two simultaneous equations are

$$x + \quad y = 500$$

$$0.8x + 0.3y = 300$$

$$x = \frac{\begin{vmatrix} c & b \\ 500 & 1 \\ 300 & 0.3 \end{vmatrix}}{\begin{vmatrix} a & b \\ 1 & 1 \\ 0.8 & 0.3 \end{vmatrix}} = \frac{-150}{-0.5} = 300 \text{ g}$$

$$y = \frac{\begin{vmatrix} a & c \\ 1 & 500 \\ 0.8 & 300 \end{vmatrix}}{\text{Denominator}} = \frac{-100}{-0.5} = 200 \text{ g}$$

Another example is concerned with investments.

EXAMPLE 7. Mr. Jones has $20,000 to invest. He can earn 9.5% on a long-term investment and 5.25% on a savings account. He wishes to keep some money in the savings account for ready access. What amount should he place in each investment to earn a total of $1500 annual interest?

Solution: Let

$x = $ amount invested at 9.5%

$y = $ amount invested at 5.25%

Then $x + y = 20,000$. The first investment earns $0.095x$ and the second investment earns $0.0525y$. Then

$$0.095x + 0.0525y = 1500$$

The two simultaneous equations are

$$x + \quad y = 20,000$$

$$0.095x + 0.0525y = 1500$$

$$x = \frac{\begin{vmatrix} c & b \\ 20{,}000 & 1 \\ 1{,}500 & 0.0525 \\ \hline 1 & 1 \\ 0.095 & 0.0525 \end{vmatrix}} = \frac{-450}{-0.0425} = \$10{,}588.24$$

$$y = \frac{\begin{vmatrix} a & c \\ 1 & 20{,}000 \\ 0.095 & 1{,}500 \\ \hline & \text{Denominator} \end{vmatrix}} = \frac{-400}{-0.0425} = \$9{,}411.76$$

Now we will demonstrate some problems involving third-order determinants.

EXAMPLE 8.   A triangle has a perimeter of 45 in. The longest side is 5 in longer than the second side. The second side is 10 in longer than the shortest side. Solve for the lengths of the three sides.

Solution: Let

x = length of longest side

y = length of second side

z = length of shortest side

The equations become

$$x + y + z = 45$$
$$x - y \quad\quad = 5$$
$$y - z = 10$$

$$x = \frac{\begin{vmatrix} d & b & c \\ 45 & 1 & 1 \\ 5 & -1 & 0 \\ 10 & 1 & -1 \end{vmatrix}\begin{matrix} 45 & 1 \\ 5 & -1 \\ 10 & 1 \end{matrix}}{\begin{vmatrix} a & b & c \\ 1 & 1 & 1 \\ 1 & -1 & 0 \\ 0 & 1 & -1 \end{vmatrix}\begin{matrix} 1 & 1 \\ 1 & -1 \\ 0 & 1 \end{matrix}} = \frac{65}{3} = 21.67 \text{ in}$$

$$y = \frac{\begin{array}{ccc} a & d & c \\ 1 & 45 & 1 \\ 1 & 5 & 0 \\ 0 & 10 & -1 \end{array} \begin{array}{cc} 1 & 45 \\ 1 & 5 \\ 0 & 10 \end{array}}{\text{Denominator}} = \frac{50}{3} = 16.67 \text{ in}$$

$$z = \frac{\begin{array}{ccc} a & b & d \\ 1 & 1 & 45 \\ 1 & -1 & 5 \\ 0 & 1 & 10 \end{array} \begin{array}{cc} 1 & 1 \\ 1 & -1 \\ 0 & 1 \end{array}}{\text{Denominator}} = \frac{20}{3} = 6.667 \text{ in}$$

EXAMPLE 9. By weight one alloy is 60% copper, 30% zinc, and 10% nickel. A second alloy is 50% copper, 30% zinc, and 20% nickel. A third alloy is 30% copper and 70% nickel. How much of each must be mixed to give a resulting alloy whose composition is 40% copper, 15% zinc and 45% nickel?

Solution: Let

x = quantity of first alloy

y = quantity of second alloy

z = quantity of third alloy

The first equation shows the percent of copper, the second the percent of zinc, and the third the percent of nickel:

$$60x + 50y + 30z = 40$$

$$30x + 30y \qquad = 15$$

$$10x + 20y + 70z = 45$$

$$x = \frac{\begin{array}{ccc} d & b & c \\ 40 & 50 & 30 \\ 15 & 30 & 0 \\ 45 & 20 & 70 \end{array} \begin{array}{cc} 40 & 50 \\ 15 & 30 \\ 45 & 20 \end{array}}{\begin{array}{ccc} a & b & c \\ 60 & 50 & 30 \\ 30 & 30 & 0 \\ 10 & 20 & 70 \end{array} \begin{array}{cc} 60 & 50 \\ 30 & 30 \\ 10 & 20 \end{array}} = \frac{0}{30,000} = 0$$

$$y = \frac{\begin{array}{ccc} a & d & c \\ 60 & 40 & 30 \\ 30 & 15 & 0 \\ 10 & 45 & 70 \end{array} \begin{array}{cc} 60 & 40 \\ 30 & 15 \\ 10 & 45 \end{array}}{\text{Denominator}} = \frac{15{,}000}{30{,}000} = 0.5$$

$$z = \frac{\begin{array}{ccc} a & b & d \\ 60 & 50 & 40 \\ 30 & 30 & 15 \\ 10 & 20 & 45 \end{array} \begin{array}{cc} 60 & 50 \\ 30 & 30 \\ 10 & 20 \end{array}}{\text{Denominator}} = \frac{15{,}000}{30{,}000} = 0.5$$

One of the most important applications is in problems involving Kirchhoff's laws of electrical circuits. They state:

1. The algebraic sum of all the currents flowing toward a junction is zero.
2. The algebraic sum of all the potential source voltages is equal to the algebraic sum of all the IR drops in any loop (Malmstadt et al., 1963).

EXAMPLE 10. Determine the currents flowing through each of the three branches of the circuit shown in Fig. 8.1. The equations can be written as follows:

$$I_A + I_B + I_C = 0 \text{ (first law)}$$
$$4I_A - 10I_B + = 3$$
$$ - 10I_B + 5I_C = 6 \text{ (second law)}$$

$$I_A = \frac{\begin{array}{ccc} d & b & c \\ 0 & 1 & 1 \\ 3 & -10 & 0 \\ 6 & -10 & 5 \end{array} \begin{array}{cc} 0 & 1 \\ 3 & -10 \\ 6 & -10 \end{array}}{\begin{array}{ccc} a & b & c \\ 1 & 1 & 1 \\ 4 & -10 & 0 \\ 0 & -10 & 5 \end{array} \begin{array}{cc} 1 & 1 \\ 4 & -10 \\ 0 & -10 \end{array}} = \frac{15}{-110} = -0.1364 \text{ A}$$

$$I_B = \frac{\begin{array}{ccc} a & d & c \\ 1 & 0 & 1 \\ 4 & 3 & 0 \\ 0 & 6 & 5 \end{array} \begin{array}{cc} 1 & 0 \\ 4 & 3 \\ 0 & 6 \end{array}}{\text{Denominator}} = \frac{39}{-110} = -0.3545 \text{ A}$$

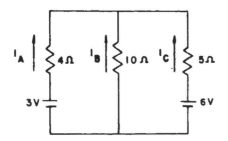

FIG. 8.1

$$I_C = \cfrac{\begin{vmatrix} a & b & d \\ 1 & 1 & 0 \\ 4 & -10 & 3 \\ 0 & -10 & 6 \end{vmatrix}\begin{matrix} 1 & 1 \\ 4 & -10 \\ 0 & -10 \end{matrix}}{\text{Denominator}} = \frac{-54}{-110} = 0.4909 \text{ A}$$

III. The Derivative

It is not the purpose of this text to instruct the student in calculus. We will only illustrate ways in which the calculator may be used to assist in the solution of calculus problems. Differential calculus is used to solve problems involving the rate of change of two quantities with respect to each other. The derivative is defined as the instantaneous rate of change of one quantity with respect to another. The derivative of y with respect to x is expressed as dy/dx.

We will review briefly the basic methods of determining the derivative (Washington, 1970; Woods and Bailey, 1928).

The first rule involves polynomials. For the function

$$y = x^n \qquad \frac{dy}{dx} = nx^{(n-1)} \tag{1}$$

EXAMPLE 11. Find the derivative of the function $y = x^3$.

Solution:

$$\frac{dy}{dx} = 3x^2$$

If the function contains several terms including a number, the number is dropped in the derivative.

EXAMPLE 12. Find the derivative of the function $y = 5x^2 + 3x + 2$.

Solution:

$$\frac{dy}{dx} = (5)(2x) + 3 = 10x + 3$$

Where a root is involved, the function may be expressed as a fractional power.

EXAMPLE 13. Find the derivative of the function $y = \sqrt{x}$. This may be written as $y = x^{1/2}$.

Solution:

$$\frac{dy}{dx} = \frac{1}{2}x^{-1/2} = \frac{1}{2\sqrt{x}}$$

Where a reciprocal is involved, it may be expressed as a negative power.

EXAMPLE 14. Find the derivative of the function $y = 1/x^2$. This may be written as $y = x^{-2}$.

Solution:

$$\frac{dy}{dx} = (-2)(x^{-3}) = \frac{-2}{x^3}$$

The second rule involves derivatives of products of functions. The rule is as follows:

$$\frac{dy}{dx} = u\frac{dv}{dx} + v\frac{du}{dx} \tag{2}$$

where

u = first function
v = second function

EXAMPLE 15. Find the derivative of the function $y = (x^3 + 3)(5 - 3x)$.

Solution: Let $u = x^3 + 3$ and $v = 5 - 3x$.

$$\frac{dy}{dx} = (x^3 + 3)(-3) + (5 - 3x)(3x^2) = -3x^3 - 3 + 15x^2 - 9x^3$$

$$= -12x^3 + 15x^2 - 3$$

The third rule involves the derivative of the quotient of two functions. The rule is as follows:

$$\frac{dy}{dx} = \frac{v \, du/dx - u \, dv/dx}{v^2} \tag{3}$$

where

u = numerator
v = denominator

EXAMPLE 16. Find the derivative of the following function:

$$y = \frac{x^2 + 2}{4 - x}$$

Solution: Let u = x^2 + 2, v = 4 − x, du/dx = 2x, dv/dx = −1. Substituting in equation (3),

$$\frac{dy}{dx} = \frac{(4 - x)(2x) - (x^2 + 2)(-1)}{(4 - x)^2}$$

$$= \frac{8x - 2x^2 + x^2 + 2}{(4 - x)^2}$$

$$= \frac{-x^2 + 8x + 2}{(4 - x)^2}$$

The fourth rule involves the derivative of powers of functions. The rule is as follows:

$$\frac{du^n}{dx} = nu^{n-1} \frac{du}{dx} \tag{4}$$

where

u = function
n = power

EXAMPLE 17. Find the derivative of the function y = $(4x^2 + 5)^3$.

Solution: Let u $= 4x^2 + 5, n = 3, du/dx = 8x$. Substituting in equation (4).

$$\frac{dy}{dx} = 3(4x^2 + 5)^2(8x) = 24x(4x^2 + 5)^2$$

IV. Applications of the Derivative

A. Slope

One of the most basic applications of the derivative is to determine the slope of a curve at a given point. The procedure is to solve for the derivative of the equation for the curve and then to substitute in the derivative the x value at which the slope is desired.

EXAMPLE 18.   For the parabola whose equation is $y = (1/2)x^2 - 3$, find the slope at the point $(1, -2.5)$. (See Fig. 8.2.)

Solution: $dy/dx = (0.5)(2x) = x$. Therefore for $x = 1$, slope $= 1$. Arctan $1 = 45°$.

B. Curvilinear Motion

A great many motion relationships involve the time rate of change. In describing an object undergoing curvilinear motion it is common practice to express the x and y coordinates of a position separately as functions

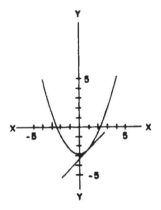

FIG. 8.2

of time. Equations given in this form are said to be in parametric form. The third variable t is called the parameter.

To find the velocity of such an object we solve for the x component of its velocity by determining dx/dt and solve for the y component of its velocity by determining dy/dt. For any given value of t, these derivatives are evaluated by substituting the value of t. This then becomes a vector problem.

$$v = \sqrt{v_x^2 + v_y^2} \tag{5}$$

$$\Theta = \arctan \frac{v_y}{v_x} \tag{6}$$

EXAMPLE 19. The motion of a particle is described by the equations $x = 10/(2t + 3)$, $y = 1/t^2$. Determine the magnitude and direction of velocity when $t = 5$. (See Fig. 8.3.)

Solution: Rearrange equations and solve for the derivatives:

$$x = 10(2t + 3)^{-1} \qquad \frac{dx}{dt} = (-10)(2t + 3)^{-2}(2) = \frac{-20}{(2t + 3)^2}$$

$$y = t^{-2} \qquad \frac{dy}{dt} = -2t^{-3} = \frac{-2}{t^3}$$

$$v_x \bigg|_{t=5} = \frac{-20}{169} = -0.1183$$

$$v_y \bigg|_{t=5} = \frac{-2}{125} = -0.016$$

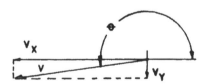

FIG. 8.3

$v = \sqrt{v_x^2 + v_y^2} = 0.1194$

$\Theta = \arctan \dfrac{-0.016}{-0.1183} = -7.7° \text{ or } 187.7°$

EXAMPLE 20. A projectile moves according to the following equations, where distances are in feet and time is in seconds: $x = 120t$, $y = 160t - 16t^2$. Solve for the magnitude and direction of v when t = 10 sec. (See Fig. 8.4.)

Solution:

$\dfrac{dx}{dt} = 120 \text{ ft/sec} = v_x$

$\dfrac{dy}{dt} = 160 - 32t$

$v_y|_{t=10} = 160 - 320 = -160 \text{ ft/sec}$

$v = \sqrt{120^2 + 160^2} = 200 \text{ ft/sec}$

$\Theta = \arctan \dfrac{-160}{120} = -53.13° \text{ or } 306.87°$

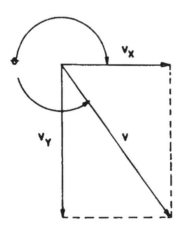

FIG. 8.4

EXAMPLE 21. A rocket follows a path given by the equation $y = x - x^3/90$ (distance in miles). If the horizontal velocity is given by $v_x = x$, find the magnitude and direction of the velocity when the rocket hits the ground (assume level terrain) (time, min).

Solution: As $y = 0$ at impact, the equation can be written as

$$x - \frac{x^3}{90} = 0$$

This reduces to $x^2 = 90$, so $x = \sqrt{90} = 9.4868$ mi. A tabulation of values and a plot of the curve are shown in Fig. 8.5. Using the value $x = 9.487$ mi/min we substitute in dy/dt to complete the solution. As

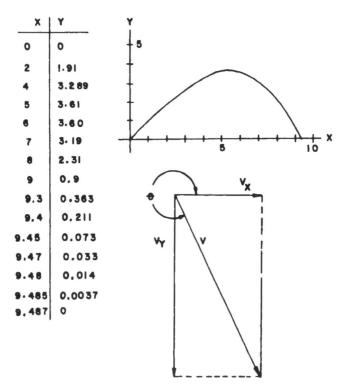

X	Y
0	0
2	1.91
4	3.289
5	3.61
6	3.60
7	3.19
8	2.31
9	0.9
9.3	0.363
9.4	0.211
9.45	0.073
9.47	0.033
9.48	0.014
9.485	0.0037
9.487	0

FIG. 8.5

both x and y change with time, both can be considered functions of time.

$$\frac{dy}{dt} = \left(1 - \frac{3x^2}{90}\right)\frac{dx}{dt}$$

$$v_y = 1 - \frac{x^2}{30}v_x = \left[1 - \frac{(9.487)^2}{30}\right]9.487$$

$$= 9.487 - \frac{(9.487)^3}{30} = -18.975 \text{ mi/min}$$

$$v = \sqrt{v_x^2 + v_y^2} = 21.21 \text{ mi/min}$$

$$\Theta = \arctan\frac{-18.975}{9.487} = -63.43° \text{ or } 296.56°$$

C. Related Rates

Any two variables which vary with respect to time and between which a relationship is known to exist can have the time rate or change of one expressed in terms of the time rate of the other. This is done by taking the derivative with respect to time of the expression relating the variables.

EXAMPLE 22. A spherical balloon is being inflated at the rate of 5 ft³/min. Find the rate of radius increase when r = 6 ft.

Solution: The formula for the volume of a sphere is $v = (4/3)\pi r^3$.

$$\frac{dv}{dt} = \left(\frac{4}{3}\right)3\pi r^2\frac{dr}{dt} = 4\pi r^2\frac{dr}{dt}$$

As $dv/dt = 5$ ft³/min, $5 = 4\pi r^2(dr/dt)$,

$$\frac{dr}{dt} = \frac{5}{4\pi r^2}$$

$$\frac{dr}{dt}\bigg|_{r=6} = \frac{5}{4\pi 36} = 0.0111 \text{ ft/min}$$

EXAMPLE 23. The electrical resistance of a certain resistor as a function of temperature is given by the equation $R = 4.00 + 0.003T^2$, where R is in ohms and T is in degrees Celsius. If the temperature is increasing

at the rate of 0.05°C/sec, how fast does the resistance change when
T = 100°C?

Solution:

$$\frac{dR}{dt} = (2)(0.003)T\frac{dT}{dt} = 0.006T\frac{dT}{dt}$$

As dT/dt = 0.05°C/sec,

$$\frac{dR}{dt}\bigg|_{DT/dt=5} = (0.006)(100)(0.05) = 0.03\ \Omega/\text{sec}$$

EXAMPLE 24. Two planes take off from an airport at 8:00 A.M. Plane
A flies north at 150 mi/hr. Plane B flies east at 200 mi/hr. How fast are
they separating at 8:30 A.M.?

Solution: Referring to Fig. 8.6, we can represent the distance traveled
by plane A by y and the distance traveled by plane B by x. By the
Pythagorean theorem

$$z^2 = x^2 + y^2$$

Taking derivatives,

$$2z\frac{dz}{dt} = 2x\frac{dx}{dt} + 2y\frac{dy}{dt}$$

$$\frac{dx}{dt} = 200\ \text{mi/hr}\ \frac{dy}{dt} = 150\ \text{mi/hr}$$

For the elapsed time,

$$x = (200\ \text{mi/hr})(0.5\ \text{hr}) = 100\ \text{mi}$$

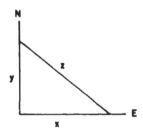

FIG. 8.6

$y = (150 \text{ mi/hr})(0.5 \text{ hr}) = 75 \text{ mi}$

$z = \sqrt{100^2 + 75^2} = 125 \text{ mi}$

$\dfrac{dz}{dt}\bigg|_{z=125} = \dfrac{(2)(100)(200) + (2)(75)(150)}{(2)(125)} = 250 \text{ mi/hr}$

D. Maxima and Minima

It is often desired to find the maximum or minimum value of a function. Two procedures are used. For the case where one equation can be used to express the relationship, the procedure is to take the derivative with respect to a variable in which it is expressed and then set the derivative equal to zero.

EXAMPLE 25.   A rectangular box is to be formed by cutting a square from each corner of a rectangular piece of metal and bending the resulting figure. If the dimensions of the sheet are 20 × 30 in, find the largest box that can be made (see Fig. 8.7) (Woods and Bailey, 1928).

Solution: The equation for the volume of the box is

$V = x(20 - 2x)(30 - 2x) = 600x - 100x^2 + 4x^3$

$\dfrac{dV}{dx} = 600 - 200x + 12x^2$

Simplifying and equating to zero,

$3x^2 - 50x + 150 = 0.$

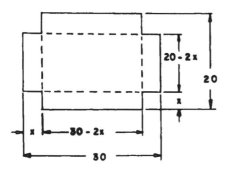

FIG. 8.7

Applying the quadratic formula,

$$x = \frac{50 \pm \sqrt{(50)^2 - (4)(3)(150)}}{(2)(3)} = 12.74 \text{ or } 3.924$$

The first answer is impossible; the second is correct. Substituting in the volume formula:

$$V = (600)(3.924) - (100)(3.924)^2 + (4)(3.924)^3 = 1056 \text{ in}^3$$

EXAMPLE 26.   The intensity of illumination from a light source at any point varies directly as the strength of the source and inversely as the square of the distance from the source. If two sources are 100 ft apart and the strength of one source is five times the strength of the other, determine at what point between them the intensity is the least, assuming that the intensity at any point is the sum of the intensities of the two sources.

Solution: Let I be the sum of the intensities and x the distance from a source of strength 5. Then

$$I = \frac{5}{x^2} + \frac{1}{(100 - x)^2} = 5x^2 + (100 - x)^{-2}$$

$$\frac{dI}{dx} = -10x^{-3} + 2(100 - x)^{-3} = \frac{-10(100 - x)^3 + 2x^3}{x^3(100 - x)^3}$$

The function will be zero if the numerator is zero:

$$2x^3 - 10(100 - x)^3 = 0 \text{ or } x^3 = 5(100 - x)^3$$

Taking the cube root of each side:

$$x = \sqrt[3]{5}(100 - x) = 100 \sqrt[3]{5} - x \sqrt[3]{5}$$

$$x(1 + \sqrt[3]{5}) = 100 \sqrt[3]{5}$$

$$x = \frac{100 \sqrt[3]{5}}{1 + \sqrt[3]{5}} = 63.099 \text{ ft}$$

If two functions can be set up based on the given information, the derivative of each function with respect to the same variable may be obtained. We can then eliminate by substitution any undesired variables or derivatives which appear in each.

EXAMPLE 27. An open top box with a square base is to be constructed using 100 ft^2 of material. What is the maximum volume that can be obtained? (See Fig. 8.8.)

Solution: Let A be the area of material used and V the volume of the box. Then

$$A = x^2 + 4xy$$

$$V = x^2y$$

$$\frac{dA}{dx} = 2x + 4x\frac{dy}{dx} + 4y = 0$$

$$\frac{dV}{dx} = 2xy + x^2\frac{dy}{dx} = 0$$

Then

$$\frac{dy}{dx} = \frac{-2xy}{x^2} = \frac{-2y}{x}$$

Substituting in the dA/dx equation:

$$\frac{dA}{dx} = 2x + (4x)\left(\frac{-2y}{x}\right) + 4y = 2x - 4y = 0$$

Then

$$y = \frac{1}{2}x$$

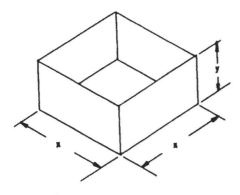

FIG. 8.8

Substituting in the area equation:

$$100 = x^2 + (4x)\frac{x}{2} = 3x^2$$

which gives

$$x = \frac{100}{3} = 5.77 \text{ ft}$$

Then

$$V = x^2y = x^2\left(\frac{x}{2}\right) = 96.23 \text{ ft}^3$$

V. Integration

A. Differentials

The derivative was defined as the rate of change of one quantity with respect to another. Integral calculus involves finding a function when the rate of change is known. Integration is the reverse of the process of differentiation. For a function $y = f(x)$ the derivative of x is defined as (Washington, 1970)

$$f'(x) = \frac{dy}{dx} \tag{7}$$

From this we can write

$$dy = f'(x) \, dx \tag{8}$$

The quantity dy is the differential of y, and dx is the differential of x.

EXAMPLE 28. Find the differential of $y = 5x^3 + x^2$.

Solution:

$$f(x) = 5x^3 + x^2$$
$$f'(x) = 15x^2 + 2x$$

Then

$$dy = (15x^2 + 2x) \, dx$$

As dy can be used to approximate Δy, this fact is useful in determining the error in a result if the data are in error.

EXAMPLE 29. The volume of a 0.5-in diameter steel ball as determined from $V = (4/3)\pi r^3$ is 0.0654 in^3. If the diameter of the ball is 0.01 in undersize, find out approximately how much the volume is reduced.

Solution:

$$V = \frac{4}{3}\pi r^3, dv = 8\pi r^2 dr$$

$$dV \cong 8\pi(0.25)^2(0.005) \cong 0.0079 \text{ in}^3$$

B. The Indefinite Integral

The basic technique of integration is the inverse of differentiation. Remember that in differentiation of an equation containing a constant, the constant disappears in the differential. Therefore we define the indefinite integral of a function f(x) for which dF(x) = f(x) as

$$\int f(x) \, dx = F(x) + C \tag{9}$$

where C is an arbitrary constant called the constant of integration. The symbol $\int$ is the integral sign, indicating that the inverse of the differential is to be found.

When we found the derivative of a power of a function, we multiplied by the power and reduced the power by 1 [equation (1)]. To find the integral we reverse this by adding 1 to the power of f(x) in f(x) dx and dividing this by the new power. The power formula for integration is stated as

$$\int u^n \, du = \frac{u^{n+1}}{n + 1} + C \tag{10}$$

EXAMPLE 30. Integrate $\int 3x \, dx$.

Solution:

$$\int 3x \, dx = 3 \int x \, dx = 3 \frac{x^2}{2} + C = \frac{3}{2}x^2 + C.$$

EXAMPLE 31. Integrate $\int (x^3 + 5x^2 + 2) \, dx$.

Solution:

$$\int (x^3 + 5x^2 + 2) \, dx = \int x^3 \, dx + 5 \int x^2 \, dx + 2 \int dx$$

$$= \frac{x^4}{4} + \frac{5x^3}{3} + 2x + C$$

EXAMPLE 32. Integrate $\int(x^2 + 2)^2(2x\ dx)$.

Solution: $n = 2$, $(x^2 + 2) = u$, $du = 2x\ dx$. Substituting in equation (10), we get

$$\int (x^2 + 2)^2(2x\ dx) = \frac{(x^2 + 2)^3}{3} + C$$

EXAMPLE 33. Integrate $2x(x^2 + 2)\ dx$.

Solution: $n = 1/2$, $u = x^2 + 2$. Since $u = x^2 + 2$, $du = 2x\ dx$. We can then write the integral as

$$\frac{1}{2} \int 2x(x^2 + 2)^{1/2}(2x\ dx) = 2(1/2)\frac{2}{3}(x^2 + 2)^{3/2} + C$$

$$= \frac{2}{3}(x^2 + 2)^{3/2} + C$$

C. The Definite Integral

When we evaluate we integral between two values of x, a and b, of f(x)

$$\int_a^b f(x)\ dx = F(b) - F(a) \tag{11}$$

where $F'(x) = f(x)$. As F(b) and F(a) both contain the constant C, we can see that C is eliminated. This is called a definite integral because the processes of integrating and evaluating produce a definite number. The numbers a and b are called the lower limit and upper limit, respectively. The value of a definite integral is found by subtracting the value of the function at the lower limit from its value at the higher limit.

EXAMPLE 34. Evaluate the integral $\int_1^3 x^3\ dx$.

Solution:

$$\int_1^3 x^3\ dx = \frac{x^4}{4}\Big|_1^3 = \frac{81}{4} - \frac{1}{4} = \frac{80}{4} = 20$$

EXAMPLE 35. Evaluate the integral $\int_0^3 x(x^2 + 9)^{-1/2}\ dx$.

Solution:

$$\int_0^3 x(x^2 + 9)^{-1/2}\, dx = (x^2 + 9)^{1/2}\, \Big|_0^3 = (9 + 9)^{1/2} - (9)^{1/2} = 1.243$$

D. Numerical Integration by the Trapezoidal Rule

Frequently, empirical data and functions cannot be directly integrated. For these cases it is possible to use numerical integration. The method we will develop is called the trapezoidal rule. It is based on the area under a curve. It will be shown how the area under the curve can be approximated by computing the area of a set of inscribed trapezoids. Referring to Fig. 8.9, the desired area is subdivided into n trapezoids of equal width Δx. If Δx is small, the approximation will be very good. From geometry we know that the area of a trapezoid equals half the product of the sum of the bases and the altitude. In this case the bases are the y coordinates and the altitudes are Δx. The sum of the areas is

$$A_T = \frac{1}{2}(y_0 + y_1)\,\Delta x + \frac{1}{2}(y_1 + y_2)\,\Delta x + \cdots + \frac{1}{2}(y_{n-1} + y_n)\,\Delta x$$

This reduces to

$$A_T = \left(\frac{1}{2}y_0 + y_1 + y_2 + \cdots + y_{n-1} + \frac{1}{2}y_n\right)\Delta x \tag{12}$$

As A_T approximates the area under the curve, it approximates the integral.

$$\int_a^b f(x)\, dx \cong A_T \tag{13}$$

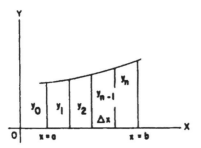

FIG. 8.9   Trapezoidal rule.

EXAMPLE 36. Approximate the value of the integral $\int_1^3 x^2\, dx$ by the trapezoidal rule. Let n = 4.

Solution:

$$\Delta x = \frac{3 - 1}{4} = 0.5$$

Now calculate values of f(x) in steps of 0.5, beginning with 1 and ending with 3:

$y_0 = (1)^2 = 1 \quad y_1 = (1 + 0.5)^2 = 2.25 \quad y_2 = (1.5 + 0.5)^2 = 4$

$y_3 = (2 + 0.5)^2 = 6.25 \quad y_4 = (3)^2 = 9$

Applying equation (13),

$$\int_1^3 x^2\, dx \cong A_T = \left(\frac{1}{2} + 2.25 + 4 + 6.25 + \frac{9}{2}\right)(0.5) \cong 8.75$$

This type of calculation is a good application for the calculator.

Calculator routine for algebraic notation:

PRESS	DISPLAY
0.5 × (1 + 2 + 2.25	
+ 4 + 6.25 + 9 ÷ 2) =	8.75

Calculator routine for reverse Polish notation:

PRESS	DISPLAY
⌐ DISP (FX) 2	
1 ENTER 2 + STO A	0.5
1.5 ⌐ x² STO + A	2.25
2 ⌐ x² STO + A	4.0
2.5 ⌐ x² STO + A	6.25
3 ⌐ x² 2 + STO + A	4.5
RCL A 0.5 ×	8.75

If smaller values of Δx are used, the accuracy is improved. By direct integration the accurate answer is 8 2/3. The percentage of error in the calculation just shown is 0.96%. If a value of $\Delta x = 0.25$ is used, the percentage of error is reduced to 0.24%. When the subject of programming is taken up in Chapter 11, it will be shown how to use smaller values of Δx conveniently.

Often it is desired to find the area of a figure bounded on one side by an irregular curved surface for which points have been established by measurement at regular intervals.

EXAMPLE 37. It is desired to find the cross-sectional area of earth to be removed at a given point in a site to be excavated. Measurements were taken at intervals of 10 feet and the following data were obtained. (See Fig. 8.10.)

x	0	10	20	30	40	50
y	10	12	11	10	9.5	10.5

We can see that $\Delta x = 10$ ft. Using the y values from the table,

$$A_T = \left(\frac{10}{2} + 12 + 11 + 10 + 9.5 + \frac{10.5}{2}\right) 10 = 527.5 \text{ ft}^2$$

Therefore the area is about 527.5 ft².

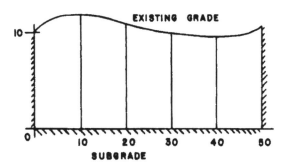

FIG. 8.10

VI. Applications of Integration

A. Areas

We have shown how to approximate the area under a curve by numerical integration. We will now demonstrate how the exact area under a curve can be determined by direct integration. In Example 36 we approximated the value of the integral $\int_1^3 x^2\,dx$. This would be the area under the curve $y = x^2$ bounded by the x axis and the lines $x = 1$ and $x = 3$. (See Fig. 8.11.) Imagine this area to be composed of an infinite number of vertical rectangular elements of widths dx and height y. Substituting the value $y = x^2$, the area of each element is $x^2\,dx$ and the area $= \int_1^3 x^2\,dx$. Following the rules of integration,

$$\int_1^3 x^2\,dx = \frac{x^3}{3}\bigg|_1^3 = \frac{27}{3} - \frac{1}{3} = \frac{26}{3} = 8\ 2/3$$

EXAMPLE 38.   Find by direct integration the area bounded by the curve $y = 9 - x^2$ and the line $y = 3 - x$, as shown in Fig. 8.12.

Solution: We can see that the curve and the line intersect at the points $(-2, 5)$ and $(3, 0)$. Again we can imagine the area to be composed of an infinite number of vertical elements whose height is $y_1 - y_2$ and whose width is dx. y_1 represents the value expressed by the equation

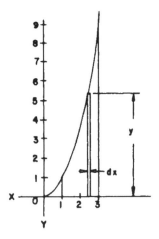

FIG. 8.11   Area under curve $y = x^2$.

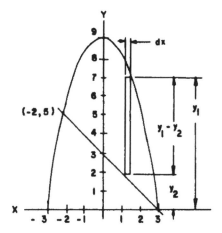

FIG. 8.12

$y = 9 - x^2$ and y_2 represents the value expressed by the equation $y = 3 - x$. Therefore

$$y_1 - y_2 = 9 - x^2 - (3 - x) = 6 - x^2 + x$$

The equation for the area is

$$A = \int_{-2}^{3} (6 - x^2 + x) \, dx = 6x - \frac{x^3}{3} + \frac{x^2}{2} \Big|_{-2}^{3}$$

$$= (18 - 9 + 4.5) - \left(-12 + \frac{8}{3} + 2\right) = 20.8333$$

B. Volumes

To find the volume of any figure generated by rotating the area bounded by a curve around an axis the procedure is as follows:

Referring to Fig. 8.13, consider the volume to be the sum of an infinite number of cylindrical elements of radius x and width dy. As the volume of a cylinder is equal to $\pi r^2 h$, the volume of the element is equal to $\pi x^2 \, dy$. Therefore the volume of the solid is

$$V = \pi \int_{a}^{b} [f(x)]^2 \, dy \qquad (14)$$

EXAMPLE 39. Determine the volume generated by rotating the area bounded by the curve $y = 3 - x^2$ and the x axis.

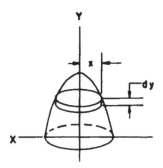

FIG. 8.13 Integration of volumes.

Solution: It can be seen that the limits of this figure are y = 3 and y = 0. The volume equation then becomes

$$V = \pi \int_0^3 x^2 \, dy = \pi \int_0^3 (3 - y) \, dy$$

$$= \pi \left(3y - \frac{y^2}{2} \right) \Big|_0^3 = \pi(9 - 4.5) = 4.5\pi$$

C. Centroids and Center of Gravity

It should be remembered that, as stated in Chapter 6, the centroids of an area are determined by the following equations:

$$\bar{x} = \frac{\Sigma Ax}{\Sigma A} \tag{15}$$

$$\bar{y} = \frac{\Sigma Ay}{\Sigma A} \tag{16}$$

EXAMPLE 40. Find $\bar{x}$ and $\bar{y}$ by integration for the right triangle illustrated in Fig. 8.14.

Solution: It can be seen that this triangle is bounded by the line whose equation is y = 2x and the line x = 3. To solve for $\bar{x}$, consider the triangle to be made up of an infinite number of vertical rectangular elements of width dx and height y. Each of these elements is at a distance x from the y axis.

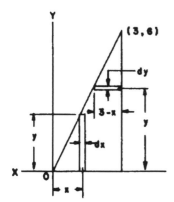

FIG. 8.14

$$\bar{x} = \frac{\int_0^3 xy \, dx}{\int_0^3 y \, dx} = \frac{\int_0^3 2x^2 \, dx}{\int_0^3 2x \, dx}$$

$$= \frac{2x^3/3}{x^2} \Big|_0^3 = \frac{18}{9} = 2$$

To solve for y, consider the triangle to be made up of an infinite number of horizontal elements of width dy and length $3 - x$. Each of these elements is at a distance y from the x axis.

$$\bar{y} = \frac{\int_0^6 y(3 - x) \, dy}{\int_0^6 (3 - x) \, dy} = \frac{\int_0^6 (3y - y^2/2) \, dy}{\int_0^6 (3 - y/2) \, dy}$$

$$= \frac{3y^2/2 - y^3 6}{3y - y^2/4} \Big|_0^6 = \frac{(3)(18) - 36}{18 - 9} = \frac{18}{9} = 2$$

In Chapter 6 it was also shown that the center of gravity of the volume of any circular solid is

$$\bar{x} = \frac{\Sigma Vx}{\Sigma V} \quad \text{or} \quad \bar{y} = \frac{\Sigma Vy}{\Sigma V} \tag{17}$$

depending on which axis is the center of the solid.

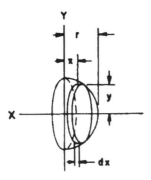

FIG. 8.15

EXAMPLE 41. Find the center of gravity of a hemisphere by integration. (See Fig. 8.15.)

Solution: To solve for $\bar{x}$, consider the hemisphere to be made up of an infinite number of cylindrical elements of radius y and height dx. Each of these elements is at a distance x from the y axis.

$$\bar{x} = \frac{\displaystyle\int_0^r x\pi y^2 \, dx}{\displaystyle\int_0^r \pi y^2 \, dx} = \frac{\displaystyle\int_0^r \pi x(r^2 - x^2) \, dx}{\int_0^r \pi(r^2 - x^2) \, dx}$$

$$= \frac{x^2 r^2/2 - x^4/4}{x^2 r^2 - x^3/3} \bigg|_0^r = \frac{r^4/2 - r^4/4}{r^3 - r^3/3} = \frac{r^4/4}{2r^3/3} = (3/8)r$$

The examples of calculus applications in this chapter are basic. They should, however, be adequate to demonstrate the usefulness of the calculator in their solution. If you encounter a problem that is difficult to integrate, you can always obtain a good approximation by using numerical integration.

Exercise 8.1

Solve the following simultaneous equations, using determinants, and check by substitution:

8.1-1 $5x + 2y = 11$

 $8x - y = 26$

8.1-2 $3x - 2y = 9$

$7x + 3y = 44$

8.1-3 $2x + 7y = 20$

$5x - 3y = 15$

8.1-4 $2x + 3y = 19$

$-3x + 5y = 41$

8.1-5 $3x + y = 10$

$4x - 2y = 6$

Exercise 8.2

Solve the following simultaneous equations, using determinants and check by substitution:

8.2-1 $x + y + z = 4$

$2x + 5y - 3z = 19$

$5x - 3y + 5z = 4$

8.2-2 $2x - 3y + z = 21$

$x + 5y - 2z = -14$

$5x + 2y + z = 21$

8.2-3 $2x + 3y + z = 5$

$x - 2y + 3z = -6$

$3x - 3y + z = 0$

8.2-4 $3x + 2y - z = 7$

$2x + y = 10$

$y + 2z = 8$

8.2-5 $x + y + z = 0$

$x + 2y = 5$

$3y - z = 2$

Exercise 8.3

8.3-1 $25,000 is invested, part at 9.5%, part at 8%, and part at 5.25%, yielding an annual interest of $2,000. The interest from the 9.5% investment yields $300 annually more than the other two combined. How much money is invested at each rate?

8.3-2 How many cubic centimeters of a 25% solution of hydrochloric acid must be mixed with a 5% solution to obtain 200 cm³ of a 10% solution?

8.3-3 Determine by Kirchhoff's laws the current flowing in each branch of the circuit shown in Fig. 8.16.

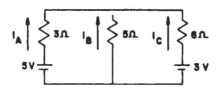

FIG. 8.16

8.3-4 A boat traveled 20 miles downstream in 3 hrs and made the return trip in 4 hrs. Determine the speeds of the boat and the current.

8.3-5 Determine by Kirchhoff's laws the current flowing in each branch of the circuit shown in Fig. 8.17.

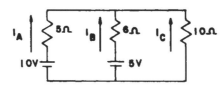

FIG. 8.17

Exercise 8.4

8.4-1 For the equation $y = x^3/2$ solve for the slope at $x = 4$.

8.4-2 For the equation $y = 3x^2$ solve for the slope at $x = 2$.

8.4-3 Find the magnitude and direction of the velocity of a projectile which moves according to the equations $x = 100t$ and $y = -16t^2$ at $t = 3$ sec, where distances are in feet and V is in ft/sec.

8.4-4   A particle moves along the curve $y = 5 x^3$ with a constant x component of -15 ft/sec. Find the magnitude and direction of the velocity at $(-2, -40)$.

8.4-5   A rocket follows a path given by the equation $y = x - x^3/50$ miles. If the horizontal velocity V_x is equal to x, find the magnitude and direction of the velocity when the rocket strikes the ground if t is in minutes.

8.4-6   The relation between voltage E and the current I that it produces in a wire of diameter d is $E = 0.124I/d^2$, where d is in inches. If the current increases at the rate of 0.03 A/sec in a 0.1 in diameter wire, what is the rate at which the voltage is increasing?

8.4-7   Boyle's law for gases states that when the temperature is constant, pressure varies inversely with volume. If a gas occupies 1000 cm^3 when the pressure is 1 atmosphere and the volume is decreasing at the rate of 25 cm^3/min, how fast is the pressure changing when the volume is 700 cm^3?

8.4-8   Design the most economical cylindrical container (closed at both ends) to hold a volume of 1 liter.

8.4-9   An area is made up of a rectangle with a semicircle at each end. If the perimeter of the figure is 440 yd, what will be the dimensions of the figure for the area of the rectangle to be a maximum?

8.4-10   The strength S of a rectangular beam is proportional to the product of its width b and the square of its depth d. Find the dimensions of the strongest beam that can be cut from a 12-in-diameter log.

Exercise 8.5

For problems 1 to 3, find the differential of each of the given functions.

8.5-1 $y = x^4 + 2x$

8.5-2 $y = (x^2 - 2)^3$

8.5-3 $y = 1/x^3$

8.5-4   The voltage of a certain thermocouple as a function of temperature is given by the equation $E = 6.2T + 0.0002T^3$. What is the approximate change in voltage if the temperature changes from 100 to 105°C?

8.5-5   If the side of a cube measures 3 in with an error of 0.1 in, what is the approximate error in the original calculation of volume?

For problems 6 to 8, evaluate the given integrals.

8.5-6 $\int_1^3 x^{3/2} \, dx$

8.5-7 $\int_2^6 (1 - \sqrt{x})^2 \, dx$

8.5-8 $\int_2^5 \sqrt[3]{4x - 3} \, dx$

8.5-9   Use the trapezoidal rule to approximate the value of the following integral:

$$\int_1^5 \frac{1}{\sqrt{x}} \, dx \qquad n = 4$$

8.5-10   A tract of land is bounded on one side by a bend in a stream and on the other side by a fence. Measurements perpendicular to the fence and to the stream were taken at 50-ft intervals and are tabulated below. Determine the approximate area of the tract in acres.

x (ft)	0	50	100	150	200	250	300
y (ft)	200	170	150	120	90	40	0

Exercise 8.6

8.6-1   Find the area bounded by the curve $y = x^2/4$ and the line $x = 4$.

8.6-2   Find the area in the first quadrant bounded by the curve $y = 8 - \left(\frac{1}{2}\right) x^2$ and the x axis.

8.6-3   Find the area bounded by the line whose equation is $y = 1 + x$ and the curve whose equation is $y = x^2/2$.

8.6-4   Derive the formula for the volume of a cone whose base has a radius r and whose height is h.

8.6-5   A casting whose form is described by rotating the curve $y - x^2/4 + 2$ around the x axis has a length extending from $x = 0$ to $x = 3$. All dimensions are in inches. What is the volume of the casting?

8.6-6   Solve for $\bar{y}$ the area bounded by the curve $y = x^2/2$ and the line $y = 4.5$.

8.6-7   Find the value of $\bar{y}$ for the triangle bounded by the line $y = 3 - x$ and the x and y axes.

8.6-8   Find $\bar{x}$ and $\bar{y}$ for the area bounded by the curve $y^2 = 4x$ and the line $x = 4$.

8.6-9   A frustum of a cone has a height of 4 in and has a 6-in-diameter at the base and 2-in-diameter at the top. Locate its center of gravity with respect to the small end.

8.6-10 Find the center of gravity of the volume generated by revolving the curve $y = 9 - x^2$ around the y axis.

9 · Principal and Interest

I. Compound Interest Problems

The principles involved in calculating interest should be understood by anyone who deals with investment. They also are applied to the determination of depreciation. The accumulation from compound interest is calculated by the following equation:

$$A = P \left(1 + \frac{r}{c}\right)^{ct} \tag{1}$$

where

    A = accumulated sum from compound interest
    t = term of investment, years
    c = number of interest periods per year
    n = ct = total number of interest periods
    r = annual rate of interest

EXAMPLE 1. Calculate the accumulation from $5000 invested at 8.5%, compounded quarterly for 10 years.

$$A = P\left(1 + \frac{r}{c}\right)^{ct} = (5000)\left(1 + \frac{0.085}{4}\right)^{(4)(10)} = \$11,594.52$$

Calculator routine for algebraic notation:

ENTER	PRESS	DISPLAY
5000	× (	
1	+	
0.085	÷	
4	) y^x (	
4	×	
10	) =	11,594.52

Calculator routine for reverse Polish notation:

PRESS	DISPLAY
⌐ DISP (FX) 2	
0.085 ENTER 4 ÷ 1 +	
4 ENTER 10 × y^x 5000 ×	11,594.52

Banks usually compound interest on passbook savings accounts daily on the basis of 366 interest periods per year.

EXAMPLE 2. Calculate the interest earned in 200 days from a deposit of $1,500 invested at 5.5% compounded daily.

$$A = P\left(1 + \frac{r}{c}\right)^{n} = 1500\left(1 + \frac{0.055}{366}\right)^{200} = \$1,545.76$$

Interest = $1,545.76 − $1,500 = $45.76

II. Payment Schedules

There are many times when a knowledge of payment schedules can be of value in the management of personal business affairs. The following formula is used to calculate the payment schedules (Texas Instruments, 1977b).

$$\text{Pmt} = \text{PV} \left[\frac{i}{1 - (1 + i)^{-n}} \right] \tag{2}$$

where

Pmt = payment per month
PV = present value
i   = interest rate per month
n   = total number of monthly payments

EXAMPLE 3. What will be the monthly payments on a 30-year home mortgage of $25,000.00 if the annual rate is 9%?

$$n = 30 \times 12 = 360$$

Applying equation (2):

$$\text{Pmt} = 25,000 \left[\frac{0.09/12}{1 - (1 + 0.09/12)^{-360}} \right] = \$201.16$$

Calculator routine for algebraic notation:

PRESS	DISPLAY
0.09 ÷ 12 = <u>STO</u> <u>ALPHA</u> A =	0.0075
÷ (1 − (1 + <u>ALPHA</u> A) y^x (−)	
360) × 25000 =	201.16

Calculator routine for reverse Polish notation:

PRESS	DISPLAY
↰ <u>DISP</u> (<u>FX</u>) 2	
0.09 <u>ENTER</u> 12 ÷ <u>STO</u> A	
1 <u>ENTER</u> 1 <u>ENTER</u> <u>RCL</u> A +	1.0075
360 <u>+/−</u> y^x − ÷	0.008
25000 ×	201.16

III. Remaining Balances

When a person is paying interest on a mortgage and must pay off the mortgage when the property is sold, it becomes necessary to determine the remaining balance owed. In this case the following formula applies:

$$Bal_K = Pmt \left[\frac{1 - (1 + i)^{K-n}}{i} \right] \tag{3}$$

where

n = total number of periods, months
i = interest rate per month
K = current period
Bal_K = balance after K-th payment

For tax purposes it is necessary each year to determine the amount of interest paid. The amount paid in any period of time is equal to the remaining balance at the end of the period minus the remaining balance at the beginning of the period. The interest paid for a given period is equal to the total of payments for the period minus the amount paid off during the period.

EXAMPLE 4. The person with the mortgage described in Example 3 decides to sell his home after 5 years. Calculate the remaining balance owed on the mortgage.

$K = 5 \times 12 = 60$

$K - n = 60 - 360 = -300$

Applying equation (3):

$$Bal_K = 201.16 \left[\frac{1 - (1 + 0.09/12)^{-300}}{0.09/12} \right] = \$23,970.55$$

Calculator routine for algebraic notation:

PRESS	DISPLAY
0.09 ÷ 12 = STO ALPHA A =	0.0075
((1 − (1 + ALPHA A) yˣ (−)	
300 + ALPHA A) × 201.06 =	23,970.55

Calculator routine for reverse Polish notation:

PRESS	DISPLAY
0.09 <u>ENTER</u> 12 + <u>STO</u> A	
1 + <u>ENTER</u> 300 <u>+/−</u> y^x	0.1063
<u>+/−</u> 1 + <u>RCL</u> A +	119.16
201.16 ×	23,970.55

EXAMPLE 5. The person with the mortgage mentioned in Examples 3 and 4 wishes to determine the amount of interest that he paid during the last year that he owned the property. First it will be necessary to determine the balance owed at the beginning of the fifth year.

For the fourth year,

$$K = 4 \times 12 = 48$$

$$K - n = 48 - 360 = -312$$

Applying equation (3):

$$Bal_K = 201.16 \left[\frac{1 - (1 + 0.09/12)^{-312}}{0.09/12} \right] = \$24{,}215.04$$

Amount paid off during the fifth year:

$$\$24{,}215.04 - \$23{,}970.55 = \$244.49$$

Payments made during fifth year:

$$12 \times \$201.16 = \$2413.92$$

Interest paid during fifth year:

$$\$2413.92 - \$244.49 = \$2169.43$$

It can be seen that interest payments are very high during the first few years of mortgage payments.

The subject of principal and interest was presented primarily to develop skill in using the calculator to work with personal finances.

Exercise 9.1

9.1-1 How much will a \$5000.00 certificate of deposit be worth after 6 years if it earns 7.75% compounded quarterly?

9.1-2 The amount of \$2000.00 is deposited in a passbook savings account which earns 5.5% compounded daily. How much does it earn in 175 days?

9.1-3 How much more will $10,000.00 earn in 183 days at 9% compounded daily than it would earn at the same rate compounded quarterly?

9.1-4 A person is purchasing a $60,000.00 home on the following terms: Downpayment is 20% with the balance covered by a 30-year mortgage at 9.5%. What will the monthly payments be?

9.1-5 How much interest does one pay for the first year on the mortgage in the preceding problem?

9.1-6 A person has paid for 15 years on a 30-year $20,000.00 mortgage with an interest rate of 8.5%. If he sells the property, how much does he still owe?

10 · Statistical Problems

I. Definition of Terms

Statistics is the science of analyzing numerical data. There are many applications of statistics in science and industry. The determination of product liability and quality control are two of the most important ones.

We will define a few statistical terms:

1. The mean is the average of all the data.
2. Deviation is the difference between any item of data and the mean.
3. The standard deviation σ is a means of predicting the distribution of a set of data. There are two ways of determining the value of σ. The first, which we shall designate σ_s, is used for analyzing small samples. The second, which we shall designate σ_p, is used for analyzing large populations of data.

$$\sigma_s = \sqrt{\frac{\Sigma x^2}{n - 1}} \tag{1}$$

$$\sigma_p = \sqrt{\frac{\Sigma x^2}{n}} \tag{2}$$

where

x = deviation of an item of data from the mean
n = number of items of data

The distribution of any random set of data will plot a normal, or bell-shaped, curve if the population is large enough. A typical normal curve is shown in Fig. 10.1. The horizontal line is the asymptote of the curve and represents the number of items occurring at each frequency. The mean is the vertical center line. It has been mathematically established that 34.13% of the data will fall within one standard deviation of the mean, 47.73% will fall within two standard deviations of the mean, 49.86% will fall within three standard deviations of the mean. These figures are multiplied by 2 to give the total percentage falling on both sides of the mean. Three standard deviations will include 99.72% of the data (Rickmers and Todd, 1967).

Table 10.1 illustrates how the standard deviation is derived.

II. Statistical Functions on the Calculator

All three calculators have routines which greatly simplify the determination of mean and standard deviation from a given set of data (Texas Instruments, 1989; Casio, Inc; Hewlett Packard, 1990). As the key sequences for statistical calculations for each of these calculators are quite different, a detailed explanation follows. We will explain these as the

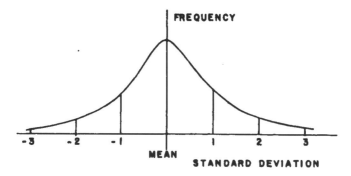

FIG. 10.1 Normal curve.

TABLE 10.1 Finding the Standard Deviation in Test Scores

No.	Score	Deviation from Mean, x	x^2
1	71	0.5	0.25
2	65	− 5.5	30.25
3	41	−29.5	870.25
4	80	9.5	90.25
5	73	2.5	6.25
6	78	7.5	56.25
7	71	0.5	0.25
8	69	− 1.5	2.25
9	71	0.5	0.25
10	79	8.5	72.25
11	75	4.5	20.25
12	73	2.5	6.25
Total	846		1,155.00
Mean	70.5		96.25

$$\sigma_1 = \sqrt{\frac{\Sigma x^2}{n-1}} = \frac{1155}{11} = 10.247$$

example problems are presented. Example 1 illustrates the procedures for unpaired data. Before beginning statistical calculations, it is necessary to clear the data registers.

EXAMPLE 1. From the data in Table 10.1 calculate the mean and σ_1, using the calculator statistical routine.

For the TI-68 press "2nd CS" (located in the first column above "Σ +"). Display shows "Clr YN." You can cancel by responding "N" or proceed with clearing by responding "Y."

Entry:

1. Type the number that is the data part.
2. Press "Σ +." If the same data occurs several times, press "Σ +" for each occurrence.
3. If you wish to remove the data point just entered, press "INV Σ +" directly after entry.

A one variable set can yield the following results:

Result	Key sequence
mean of x	2nd $\bar{x}$ =
population std. deviation	2nd σ xn =

Calculator routine for the TI-68:

PRESS	DISPLAY
71 Σ +	
65 Σ +	
41 Σ +	
80 Σ +	
73 Σ +	
78 Σ +	
71 Σ +	
69 Σ +	
71 Σ +	
69 Σ +	
75 Σ +	
73 Σ +	
2nd $\bar{x}$ =	70.5 mean
2nd σ xn =	10.247 std. deviation

For the FX-7000GA the explanation of the key sequence may be confused by the fact that certain functions are performed by pressing keys that are labeled for other functions. Statistical functions are performed in the "SDI" mode. (Press "MODE X," the multiplication key).

Clear statistical memories by pressing "SHIFT AC," "EXE."
Entry:

1. Type the number that is the data point.
2. Press "$\sqrt{\ }$ " key. Multiple data of the same value may be input by repeating "$\sqrt{\ }$."
3. If you wish to clear the data point just entered, press "x^y."

Entry:

Result	Key sequence
mean of x	SHIFT x̄ EXE
population std. deviation	SHFIT σ n EXE

Calculator routine for the FX-7000GA:

PRESS	DISPLAY
Mode x	
SHIFT AC EXE	
71 $\sqrt{\ }$	
65 $\sqrt{\ }$	
41 $\sqrt{\ }$	
80 $\sqrt{\ }$	
73 $\sqrt{\ }$	
78 $\sqrt{\ }$	
71 $\sqrt{\ }$	
69 $\sqrt{\ }$	
71 $\sqrt{\ }$	
71 $\sqrt{\ }$	
79 $\sqrt{\ }$	
75 $\sqrt{\ }$	
73 $\sqrt{\ }$	
SHIFT x̄ EXE	70.5 mean
SHIFT σ n EXE	10.47 std. deviation

For the HP-32SII the statistical register is cleared by pressing "⇥ CLEAR (Σ)."
Entry:

1. Type the number that is the data point.
2. Press "Σ +." If the same data point occurs several times, press "Σ +" for each time.
3. If you wish to remove the data point just enetered, press "⇥ Σ −."

To determine the results after data entry: Proceed as follows:

1. To obtain the mean, ⌐ x̄, ȳ. A menu appears:

   x̄ ȳ xw

   Press the key below x̄.
2. To obtain the standard deviation, press ⌐ s, σ then (s x) in the menu.

Calculator routine for the HP-32SII:

PRESS	DISPLAY
71 Σ +	
65 Σ +	
41 Σ +	
80 Σ +	
73 Σ +	
78 Σ +	
71 Σ +	
69 Σ +	
71 Σ +	
79 Σ +	
75 Σ +	
73 Σ +	
⌐ x̄, ȳ (x̄)	70.5 Mean
⌐ s, σ (s x)	10.25 Std. deviation

III. Linear Regression

Linear regression is a statistical method for finding a straight line that best fits a set of data points, thus providing a graph of the relationship between two variables. Any measurement of data, because of human error or other factors, is likely to produce some points that will not fit perfectly on a striaght line. The measure of how well a given set of data fits a straight line is called the correlation coefficient. The value will fall between 1 and −1, with ±1 being a perfect fit and 0 being no fit. Both types of calculator have routines for determining the correlation coefficient, the slope of the line, and the y intercept of the line. This makes it possible to construct a graph without having to plot all the points.

Knowing the slope and y intercept, we can write the equation for the line:

$$y = Ax + B \tag{3}$$

where

A = slope
B = y intercept

EXAMPLE 2.   In a physics experiment with a convex lens, readings are taken of the object distance and the corresponding image distance for a number of different positions of the object with respect to the lens. Table 10.2 gives the reciprocals of the object and image distances. With the calculator, determine the correlation coefficient, the slope of the line, and the y intercept. The graph of the plotted points is shown in Fig. 10.2.

TABLE 10.2   Reciprocals of Object and Image Distances

No.	$1/d_o$	$1/d_i$
1	0.015	0.05
2	0.021	0.043
3	0.029	0.036
4	0.034	0.030
5	0.043	0.022

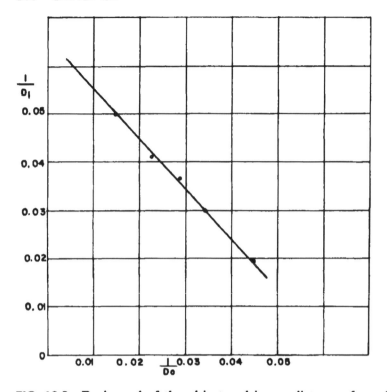

FIG. 10.2 Reciprocal of the object and image distances for a 15-cm focal length lens.

Here again, procedures for each calculator will be explained as the problems are presented. For the TI-68:

1. Type the number for the x-value, press "," and type the y-value.
2. Press "$\Sigma +$."
3. If you wish to remove the data point just entered press "INV $\Sigma +$" directly after entry.

Results are obtained as follows:

Result	Key sequence
correlation coefficient	3rd COR =
slope	3rd SLP =
y-intercept	3rd ITC =

Calculator routine for the TI-68:

PRESS	DISPLAY
0.015, 0.05 Σ +	
0.021, 0.043 Σ +	
0.029, 0.036 Σ +	
0.034, 0.03 Σ +	
0.043, 0.022 Σ +	
3rd COR =	−0.9987
3rd SLP =	−0.996
3rd ITC =	0.0645

For the FX-7000GA data entry is as follows:

1. Type the number of the x-value, press "SHIFT" "," and type the y-value
2. Press "$\sqrt{}$."
3. If you wish to remove the data just entered, enter the correct value after pressing "AC."

Results are obtained as follows:

Result	Key sequence
correlation coefficient	SHIFT r EXE
slope	SHIFT B EXE
y-intercept	SHIFT A EXE

Calculator routine for the FX-7000GA:

PRESS	DISPLAY
0.015 SHIFT , 0.05 $\sqrt{}$	
0.021 SHIFT , 0.043 $\sqrt{}$	
0.029 SHIFT , 0.036 $\sqrt{}$	
0.034 SHIFT , 0.03 $\sqrt{}$	

0.043 <u>SHIFT</u> , 0.022 $\sqrt{}$

<u>SHIFT r EXE</u>	−0.999 correlation
<u>SHIFT B EXE</u>	−0.996 slope
<u>SHIFT A EXE</u>	0.064 y-intercept

For the HP-32SII data entry is as follows:

1. Type the number of the y-value, press "ENTER" and type the x-value. (Note that this is the reverse order from the other two calculators).
2. Press "Σ +."

Results are obtained as follows:

Result	Key sequence
correlation coefficient	┌• L.R. (r)
slope	┌• L.R. (m)
y-intercept	┌• L.R. (b)

Calculator routine for the HP-32SII:

PRESS	DISPLAY
0.05 <u>ENTER</u> 0.015 Σ +	
0.043 <u>ENTER</u> 0.021 Σ +	
0.036 <u>ENTER</u> 0.029 Σ +	
0.03 <u>ENTER</u> 0.034 Σ +	
0.022 <u>ENTER</u> 0.043 Σ +	
┌• L.R. (r)	−1.00 correlation
┌• L.R. (m)	−1.00 slope
┌• L.R. (b)	0.06 y-intercept

IV. Trend Line Analysis

The value of the linear regression routine is that it enables you to make predictions about points for which you have no data. After you have

TABLE 10.3   Calibration Figures for
Pressure Transducer

Force, lb	Strain indicator, μin
1000	1224
2000	1385
3000	1545
4000	1700

entered your data in the calculator and have found that they produce a reasonably good correlation, you can with confidence ask the calculator to give you corresponding x and y values for any other points that might fall on the line.

EXAMPLE 3.   A calibration test was run between the force applied to a pressure transducer and the corresponding readings on a strain indicator. The data obtained are recorded in Table 10.3. A graph of the data is shown in Fig. 10.3. Determine the correlation coefficient. If the force is 3500 lb, what should be the strain indicator reading? If the strain indicator reading is 2500 μin, what is the corresponding force?

For the TI-68, if you wish to find the y-value corresponding to a given x-value, press "3rd y'" then the given x-value and " = ." If you wish to find the x-value corresponding to a given y-value, press "3rd x'" then the given value and " = ."

Calculator routine for the TI-68:

PRESS	DISPLAY
2nd <u>CS</u> (Y)	
1000, 1224 Σ +	
2000, 1385 Σ +	
3000, 1545 Σ +	
4000, 1700 Σ +	

<u>3rd COR</u> =   1

<u>3rd SLP</u> =   0.1588

<u>3rd ITC</u> =   1066.5

3rd y' 3500 =   1622.3   microinch

3rd x' 2500 =   9027.0   lb force

For the FX-7000GA, if you wish to find the y-value corresponding to a given x-value, type the x-value followed by "SHIFT ŷ EXE." If you wish to find the x-value corresponding to a given y-value, type the y-value followed by "SHIFT x̂ EXE."

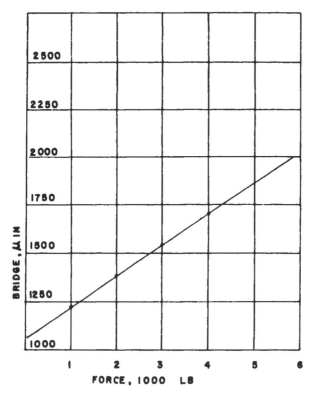

FIG. 10.3 Force vs. transducer reading.

Calculator routine for the FX-7000GA:

PRESS	DISPLAY	
MODE +		
SHIFT AC EXE		
1000 SHIFT , 1224 $\sqrt{}$		
2000 SHIFT , 1385 $\sqrt{}$		
3000 SHIFT , 1545 $\sqrt{}$		
4000 SHIFT , 1700 $\sqrt{}$		
SHIFT r EXE	1.00	correlation
SHFIT A EXE	1066.5	y-intercept
SHIFT B EXE	0.159	slope
3500 SHIFT $\hat{y}$ EXE	1622	microinches
2500 SHIFT $\hat{x}$ EXE	9027	lb force

For the HP-32SII, if you wish to find the y-value corresponding to a given x-value, type the x-value followed by ΓL.R. ($\hat{y}$)." If you wish to find the x-value corresponding to a given y-value, type the y-value followed by " ΓSTAT (LR) ($\hat{x}$)."

Calculator routine for the HP-32SII (y-values first):

PRESS	DISPLAY
ΓCLEAR (Σ)	
1224 ENTER 1000 Σ +	
1385 ENTER 2000 Σ +	
1545 ENTER 3000 Σ +	
1700 ENTER 4000 Σ +	
ΓL.R. ($\underline{r}$)	1.00 correlation
ΓL.R. ($\underline{\hat{y}}$)	1622.00 microinches
ΓL.R. ($\underline{\hat{x}}$)	9027.00 lb force

Of course, all phenomena do not display straight-line relationships.
At this point we should inject a word of caution. If a set of data fits a
parabola, it may still give a fairly good linear correlation. Remember
that part of a parabola is nearly linear. If, after making all the entries,
you enter the low x value, ask for the corresponding y value, and find
that the answer differs substantially from the data, the relationship is
probably parabolic. If the correlation is low, the relationship is obviously
not linear.

In this chapter we have attempted to give a general view of the uses
of the calculator in statistical work. In almost any field you can think
of, there is an application for statistics in the analysis of data.

Exercise 10.1

10.1-1   A group of 50 students was given the same test. Their scores
are tabulated below. Calculate the mean and standard deviation
σ_s.

No.	1	2	3	4	5	6	7	8	9	10	11	12
Score	24	56	70	37	67	91	75	85	89	45	65	67

No.	13	14	15	16	17	18	19	20	21	22	23	24
Score	75	86	94	53	60	78	89	87	95	56	65	60

No.	25	26	27	28	29	30	31	.32	33	34	35	36
Score	75	85	88	90	56	67	77	85	89	92	99	90

No.	37	38	39	40	41	42	43	44	45	46	47	48
Score	91	93	98	79	94	92	75	97	93	77	90	92

No.	49	50
Score	73	73

10.1-2 A run of steel bars are to be cut in lengths of 100 mm ±1 mm. Random samples of six pieces are taken for inspection. Following are measurements of a sample. Calculate mean and standard deviation. If 3σ represents maximum variation in each direction, is the processs within tolerance?

No.	1	2	3	4	5	6
L, mm	100.1	100.3	99.9	99.8	99.7	100.2

10.1-3 For the following data determine the correlation coefficient, the slope, and the y intercept.

x	5	10	15	20	25	30
y	3	5	7	10	12	15

10.1-4 In an experiment the velocity of a free-falling object with respect to time was determined by electronic instrumentation. The following data were obtained. Determine the correlation coefficient, the velocity at 10 sec, and the time in second for a velocity of 2500 cm/sec.

v, cm/sec	970	1970	3000	3900	5020	5900
t, sec	1	2	3	4	5	6

10.1-5 If a gas is cooled under conditions of constant volume, pressure drops are nearly proportional to temperature drops. If the temperatures are reduced until the pressure is zero, the theoretical temperature should be absolute zero. In a lab experiment the following data were obtained. Determine the correlation coefficient and the value of absolute zero indicated by this experiment.

P, cm of Hg	100	107	115	121	129	137
T, °C	0.0	20	40	50	80	100

.

11 · Programming and Special Techniques for the TI-68

This calculator has a number of routines that are quite useful in solving problems involving repetetive calculations using the same routine. The topics that will be covered in this chapter are:

Formula programming
Integration
Simultaneous equations
Polynomials

I. Formula Programming

Before proceeding with this topic, it will be necessary to explain some of the rules that apply. A formula must be assigned a name so that it can be identified when needed. This name may consist of one, two, or three characters, the first of which must be a letter. The other characters may be letters or numbers. It is recommended that the formula name be one that may be easily associated with the operation being performed.

For instance, the formula for determining the area of a circle may be named "AC."

Next, it is necessary to explain clearing procedures. Always begin the program entry procedure by pressing "CLEAR." If a formula has already been entered, and one wishes to enter a new set of variables, press "3rd (CVS)Y."

To erase a formula, press "2nd FMLA" to identify, then press "3rd (CFV) Y." To list formulas that have been entered, respond to the "NAME ?" prompt by pressing " = ." The first formula in alphabetical order will appear. To advance to another formula, press "2nd NEXT" or "2nd BACK."

EXAMPLE 1. Write a program for finding the area of a circle from the formula:

$$A = \pi d^2/4$$

Note that where a question is asked in the "Display" column, the answer appears in the next line in the "Press" column.

Calculator routine:

PRESS	DISPLAY	COMMENT
CLEAR		Clear entry line
2nd FMLA	Name?	
ALPHA A ALPHA C =		
2nd π × ALPHA D		
x^2 ÷ 4 =	Solve YN?	
Y	d = ?	Input
4 =	12.5664	Area

For entering a second variable value the procedure is as follows:

PRESS	DISPLAY	COMMENT
3rd (CVS) Y		Clears previous
	variables	
2nd (FMLA)	Name?	
ALPHA A ALPHA C =	πd^2 ÷ 4 =	
	Solve YN?	

Y	$d = ?$	Input
7.5 =	Review YN?	
N	$AC = 44.178$	Area

EXAMPLE 2.   Write a program for the Pythagorean theorem:

$$C = \sqrt{A^2 + B^2}$$

where known sides are 3 and 4.

Calculator program:

PRESS	DISPLAY	COMMENT
<u>CLEAR</u>		
2nd <u>(FMLA)</u>	Name?	
<u>ALPHA</u> C = $\sqrt{\ }$ (<u>ALPHA</u> A x^2		
+ <u>ALPHA</u> B x^2) =	Solve YN?	
Y	$A = ?$	Input
3 =	$B = ?$	Input
4 =	5	Value, C

EXAMPLE 3.   Write a program for applying the Law of Cosines:

$$C = \sqrt{a^2 + b^2 - 2ab \cos C}$$

where the following values apply: $a = 600$, $b = 40$, $C = 45$ degrees.

Calculator program:

PRESS	DISPLAY	COMMENT
<u>CLEAR</u>		
2nd <u>(FMLA)</u>	Name?	
<u>ALPHA</u> A <u>ALPHA</u> C =		
$\sqrt{\ }$ (<u>ALPHA</u> A x^2 +		
<u>ALPHA</u> B x^2 − 2 ×		
<u>COS</u> <u>ALPHA</u> C) =	Solve YN ?	
Y	$A = ?$	Input

600 =	B = ?	Input
40 =	C = ?	Input
45 =	572.4	Answer, C

EXAMPLE 4. Write a program for application of the quadratic formula:

$$x = \frac{-b \pm \sqrt{b^2 - 4ac}}{2a}$$

and solve the equation for $x^2 - 20x - 100$.

Because the rules for programming do not permit the use of the "STORE" function within the program, it will be necessary to use a different procedure from that used when this problem was presented in Chapter 3. As there are two roots, we will write three programs. The first, Q, will evaluate the quantity under the radical. The second, Q1, will evaluate:

$$-b/2a + Q$$

The third, Q2, will evaluate:

$$-b/2a - Q$$

Calculator program for Q:

PRESS	DISPLAY	COMMENT
CLEAR		
2nd (FMLA)	Name?	
ALPHA Q =		
√ (ALPHA B x² − 4 ×		
ALPHA A × ALPHA C) =	Solve YN?	
Y	B = ?	
(−)20 =	A = ?	
1 =	C = ?	
(−) 100 =	28.284	Q

Calculator program for Q1:

PRESS	DISPLAY	COMMENT
2nd (FMLA)	Name ?	
ALPHA Q1 =		
((−) ALPHA B + ALPHA Q)		
÷ 2 × ALPHA A =	Solve YN ?	
Y (Previous entries will appear)	24.142	1st root

Calculator program for Q2:

PRESS	DISPLAY	COMMENT
2nd (FMLA)	Name ?	
ALPHA Q2 =	.	
((−)ALPHA B − ALPHA Q)		
÷ 2 × ALPHA A =	Solve YN?	
Y (Previous entries will appear)	− 4.1421	2nd root

II. Integration

The procedure for calculating an integral is similar to the one for evaluating a formula. However, you declare one of the variables to be "dX" instead of assigning it a value. Follow this procedure:

1. Be sure that the calculator is set in the decimal mode. If anlges are involved, the radian mode must be selected.
2. To integrate a function, press "2nd (FMLA)," spell the function name, and press " = ."
3. Press "SOLVE." Reply to the "x = ?" inquiry with "3rd (dX) = ."
4. Respond to "Low = ?" with the lower integration limit.
5. Respond to "Up = ?" with the upper integration limit.
6. Respond to "Intrv = ?" with the desired number of intervals. (10 is suggested. Use more if greater accuracy is required.)
7. Respond to "review YN ?" with "N." If "Error" appears in the display, press "2nd (EQU)" to show the formula with the cursor positioned at the error.

EXAMPLE 5. Integrate the expression $\int_1^3 x^2\, dX$

Calculator routine:

PRESS	DISPLAY	COMMENT
CLEAR		
2nd (FMLA)	Name ?	
ALPHA F ALPHA N1 =		
ALPHA X x²		
SOLVE	X = ?	
3rd (d X) =	Low = ?	
1 =	Up = ?	
3 =	Intrv = ?	
10 =	Review YN ?	
N	FN1 = 8.6666	

EXAMPLE 6. Integrate the expression $\int_{-2}^{3} (6 - x^2 + x)\, dX$

Calculator routine:

PRESS	DISPLAY	COMMENT
CLEAR		
2nd (FMLA)	Name ?	
ALPHA F ALPHA N2 =		
6 − ALPHA X x² + ALPHA X		
SOLVE	X = ?	
3rd (d X) =	Low = ?	
(−) 2 =	Up = ?	
3 =	Intrv = ?	
10 =	Review YN ?	
N	FN2 = 20.833	

EXAMPLE 7.   Integrate the expression $\int_2^5 3\sqrt{4X-3}\,dX$

Calculator routine:

PRESS	DISPLAY	COMMENT
<u>CLEAR</u>		
2nd <u>(FMLA)</u>	Name ?	
<u>ALPHA</u> F <u>ALPHA</u> N3 =		
3 × $\sqrt{\ }$ (4 × ALPHA X − 3)		
<u>SOLVE</u>	X = ?	
3rd <u>(d X)</u> =	Low = ?	
2 =	Up = ?	
5 =	Intrv = ?	
10 =	Review YN ?	
N	FN3 = 29.4562	

EXAMPLE 8.

Integrate the expression $\int_0^2 \left(\frac{\text{Sin } X}{X}\right) dX$

As the lower limit, 0, would cause a division by 0 and yield an error message, we will substitute 0.0001 as the lowr limit.

Calculator routine:

PRESS	DISPLAY	COMMENT
3rd <u>(DRG>)</u> <u>CLEAR</u>		Radian mode
2nd <u>(FMLA)</u>	Name ?	
<u>ALPHA</u> F <u>ALPHA</u> N4 =		
(Sin ALPHA X + ALPHA X)		
<u>SOLVE</u>	X = ?	
3rd <u>(d X)</u> =	Low = ?	
0.0001 =	Up = ?	
2 =	Intrv = ?	
10 =	Review YN ?	
N	FN4 = 1.6053	

III. Simultaneous Equations

In Chapter 8 we presented methods for solution of simultaneous equations by use of determinants. The TI-68 has a simple procedure by which solutions of up to five sets of simultaneous equations may be found. This can be very useful in the solution of circuit analysis problems. Before proceeding with this topic, we will refer to page 7-2 of the manual where their arrangement of equations in rows and columns is explained. This numbering arrangement will be referred to in the solution procedure. Follow this procedure:

1. Be sure that the calculator is set in the decimal mode.
2. Examine the equations to identify their order and coefficients.
3. Press "2nd (SIMUL)." The calculator prompts you for the number of equations in the system.
4. Respond with the number of equations. The calculator asks for "Complex YN? "
5. To respond that the system is real, press "N."
6. The calculator will then ask for row 1 coefficients. Enter in order requested.
7. Proceed with entry of coefficients in the following rows.
8. After all coefficients have been entered, the message "Review YN?" appears. Respond "N."
9. The first solution, "X1" appears. For further solutions, press "2nd (NEXT)."
10. When all solutions have been found, press "2nd (EXIT)."

EXAMPLE 9. Solve the following set of simultaneous equations:

$$2X + Y = 3$$
$$5X + 3Y = 10.$$

Calculator routine:

PRESS	DISPLAY	COMMENT
CLEAR		
2nd (SIMUL)	Equa 2-5 ?	
2	Complex YN ?	
N	a11 = ?	Enter row 1 coefficients

2 =	a12 = ?	
1 =	b1 = ?	
3 =	a21 = ?	Enter row 2 coefficients
5 =	a22 = ?	
3 =	b2 = ?	
10 =	Review YN ?	
N	X1 = −1	Value of X
2nd (NEXT)	X2 = 5	Value of Y
2nd (EXIT)		

EXAMPLE 10. Solve the following set of simultaneous equations:

$X + Y = 20,000$

$0.095X + 0.052Y = 1500$

Calculator routine:

PRESS	DISPLAY	COMMENT
CLEAR		
2nd (SIMUL)	Equa 2 − 5?	
2	Complex YN?	
N	a11 = ?	
1 =	a12 = ?	
1 =	b1 = ?	
20,000 =	a21 = ?	
0.095 =	a22 = ?	
0.0525 =	b2 = ?	
1500 =	Review YN ?	
N	X1 = 10,588.24	
2nd (NEXT)	X2 = 9,411.76	
2nd (EXIT)		

EXAMPLE 11. Solve the following set of simultaneous equations:

$$2X + 3Y + Z = 7$$
$$3X - Y - 2Z = 17$$
$$4X + 5Y + 3Z = 7$$

Calculator routine:

PRESS	DISPLAY	COMMENT
CLEAR		
2nd (SIMUL)	Equa 2-5 ?	
3	Complex YN ?	
N	a11 = ?	Row 1 coefficients
2 =	a12 = ?	
3 =	a13 = ?	
1 =	b1 = ?	
7 =	a21 = ?	Row 2 coefficients
3 =	a22 = ?	
(−)1 =	a23 = ?	
(−)2 =	b2 = ?	
17 =	a31 = ?	Row 3 coefficients
4 =	a32 = ?	
5 =	a33 = ?	
3 =	b3 = ?	
7 =	Review YN ?	
N	X1 = 3	Value of X
2nd (NEXT)	X2 = 2	Value of Y
2nd (NEXT)	X3 = −5	Value of Z
2nd (EXIT)		

EXAMPLE 12. Solve the following set of simultaneous equations:

$IA + IB + IC = 0$

$4IA - 10IB = 3$

$- 10IB + 5IC = 6$

Calculator routine:

PRESS	DISPLAY	COMMENT
CLEAR		
2nd (SIMUL)	Equa 2 - 5 ?	
3	Complex YN ?	
N	a11 = ?	Row 1 coefficients
1 =	a12 = ?	
1 =	a13 = ?	
1 =	b1 = ?	
0 =	a21 = ?	Row 2 coefficients
4 =	a22 = ?	
(−)10 =	a23 = ?	
0 =	b2 = ?	
3 =	a31 = ?	Row 3 coefficients
0 =	a32 = ?	
(−)10 =	a33 = ?	
5 =	b3 = ?	
6 =	Review YN ?	
N	X1 = −0.13636	Value of IA
2nd (NEXT)	X2 = −0.35454	Value of IB
2nd (NEXT)	X3 = 0.4909	Value of IC
2nd (EXIT)		

IV. Polynomials

One of the unique features of the TI-68 is the routine for solving polynomials. The polynomial must have the correct sequence of terms. The
three allowable orders of polynomials are as follows:

Fourth order:

$$A4X^4 + A3X^3 + A2X^2 + A1X + A0 = 0$$

Third order:

$$A3X^3 + A2X^2 + A1X + A0 = 0$$

Second order:

$$A2X^2 + A1X + A0 = 0$$

Sort the terms in correct sequence.

1. Write the highest order terms at the beginning of the polynomial.
2. Place the rest of the terms in descending sequence according to the
   order of the variables in the system.
3. Combine terms of equal order in one term.
4. Include a zero coefficient for any lower order terms that are absent
   from the polynomial.

The calculator procedure for solving polynomials is similar to that
used for simultaneous equations. The rules follow:

1. Check to be sure that the calculator is set in the decimal mode.
2. Press "2nd (POLY)." The calculator prompts you for the order of
   the polynomial.
3. Respond with the order of the polynomial, 2, 3, or 4. The calculator
   prompts you for the first coefficient.
4. Respond with the value of the coefficients as requested. Type " = "
   after each entry.
5. The prompt "Review YN ?" appears. To skip reviewing, respond
   "N." The first root appears.
6. Press " = " to review other roots.
7. If no more roots appear, press "2nd (EXIT)."

EXAMPLE 13. Find the roots of the equation:

$$X^3 + 2X^2 - 7X + 5 = 0$$

Calculator routine:

PRESS	DISPLAY
CLEAR	
2nd (POLY)	Order 2-4 ?
3	A3 = ?
1 =	A2 = ?
2 =	A1 = ?
(−) 7 =	A0 = ?
5 =	Review YN ?
N	− 4.03937
2nd (EXIT)	

EXAMPLE 14. Find the roots of the equation:

$$X^4 - 7X^3 + 12X^2 + 4X - 16 = 0$$

Calculator routine:

PRESS	DISPLAY
CLEAR	
2nd (POLY)	Order 2-4 ?
4	A4 = ?
1 =	A3 = ?
(−) 7 =	A2 = ?
12 =	A1 = ?
4 =	A0 = ?
(−) 16 =	Review YN ?
N	X1 = 2
=	X2 = 4
=	X3 = −1
2nd (EXIT)	

EXAMPLE 15. A simple solution to the quadratic equation follows: Find the roots of the equation $X^2 - 20X - 100 = 0$.

Calculator routine:

PRESS	DISPLAY
CLEAR	
2nd (POLY)	Order 2-4 ?
2	A2 = ?
1 =	A1 = ?
(−)20 =	A0 = ?
(−)100 =	Review YN ?
N	X1 = −4.142
=	X2 = 24.142
2nd (EXIT)	

In this chapter we have presented techniques for quickly solving problems which would otherwise be very time consuming. It is hoped that they can be put to good use.

12 · Programming and Special Techniques for the Casio FX-7000GA

We will cover only two topics for this calculator. They are:

1. Formula programming
2. Graphics

I. Formula Programming

First let us explain the basic rules and procedures that apply. To write a program, press "MODE 2." The following typical display appears:

sys mode : WRT
cal mode : LRI
angle   : Deg
display   : Fix 3
276 Bytes Free
Prog 0 1 2 3 4 5 6 7 8 9

The bottom line indicates the program numbers available. If a program number is in use, a dash will appear in its place in the display. The first program entered will be assigned number "0." To enter a program, press "MODE 2" "EXE." If the space assigned to the program is other than "0," use ⇨ to move to program number desired. The number will start flashing. Press "EXE." Following are rules for erasing and editing programs:

1. To erase a program, set "MODE 3," move display to program number, press "AC."
2. To erase all programs, set "MODE 3," press "SHIFT" then "DEL."
3. Program editing, set "MODE 2," move display to program number, press "EXE." With display showing the program and cursor located at the beginning, move cursor by use of the cursor arrows ⇨ or ⇦ to desired position and press "DEL" to remove character, then insert correction.

EXAMPLE 1. Write a program for finding the area of a circle, using the formula:

$$A = \pi D^2/4$$

Note that the variable is preceded by "SHIFT ? →" and followed by ":". The formula is then entered, followed by "SHIFT ◿" which is the command to print the answer.

Calculator program:

PRESS	DISPLAY	COMMENT
MODE 2 EXE		
SHIFT ? → ALPHA D:		
SHIFT π × ALPHA D		
x^2 ÷ 4 SHIFT ◿	? → D:πx D^2 ÷ 4	

Program execution is performed in MODE 1

MODE 1		
Prog 0 EXE	?	
4 EXE	12.566	Area for D = 4
Prog 0 EXE	?	
7 EXE	38.485	Area for D = 7

EXAMPLE 2. Write a program for solving the Pythagorean theorem:

$$C = \sqrt{A^2 + B^2}$$

Note here that we will be assigning program 1. Following "MODE 2" use the cursor arrow to set "1" before pressing "EXE." The number will flash on the display.

Calculator program:

PRESS	DISPLAY	COMMENT
MODE 2 ⇨ 1 EXE		
SHIFT ?→ ALPHA A :		
SHIFT ?→ ALPHA B :		
√ (ALPHA A x² +		
ALPHA B x²) SHIFT ◿	→ A:?→B:	
	√ (A² + B²)	

       Program execution:

MODE 1		
Prog 1 EXE	?	
3 EXE	?	
4 EXE	5	Answ for A = 3, B = 4
Prog 1 EXE	?	
10.65 EXE	?	
17.28 EXE	20.298	Answ for 2nd values

EXAMPLE 3. Write a program for solving the law of cosines:

$$C = \sqrt{a^2 + b^2 - 2ab \, \cos C}$$

Calculator program:

PRESS	DISPLAY	COMMENT
MODE 2 ⇨ 2 EXE		
SHIFT ? → ALPHA A:		
SHIFT ? → ALPHA B:		

SHIFT ? → <u>ALPHA</u> C:

$\sqrt{}$ (<u>ALPHA</u> A x^2 +

<u>ALPHA</u> B x^2 − 2 ×

<u>ALPHA</u> B × <u>ALPHA</u> B ×

<u>COS</u> <u>ALPHA</u> C)

SHIFT ◿

    Program execution:

<u>MODE</u> 1

<u>Prog</u> 2 <u>EXE</u>	?	
600 <u>EXE</u>	?	
40 <u>EXE</u>	?	
45 <u>EXE</u>	572.4	Answ 1st values
<u>Prog</u> 2 <u>EXE</u>	?	
200 <u>EXE</u>	?	
150 <u>EXE</u>	?	
35 <u>EXE</u>	115.546	Answ 2nd values

EXAMPLE 4. Write a program for solving the payment schedule, using the formula:

$$\text{Pmt} = A \left[\frac{I/12}{1 - (1 + (I/12))^{-N}} \right]$$

Calculator program:

PRESS	DISPLAY	COMMENT

<u>MODE</u> 2 ⇨ 3 <u>EXE</u>

SHIFT ? → <u>ALPHA</u> A:

SHIFT ? → <u>ALPHA</u> I:

SHIFT ? → <u>ALPHA</u> N:

<u>ALPHA</u> A × (<u>ALPHA</u> I/12)

$+ (1 - (1 + (\underline{ALPHA}\ I/12))$

$y^x\ (-)\ \underline{ALPHA}\ N))$

Program execution:

<u>MODE</u> 1 <u>Prog</u> 3 <u>EXE</u>	?	
25000 <u>EXE</u>	?	
0.09 <u>EXE</u>	?	
360 <u>EXE</u>	201.15	Monthly pmt
<u>MODE</u> 1 <u>Prog</u> 3 <u>EXE</u>	?	
3000 <u>EXE</u>	?	
0.105 <u>EXE</u>	?	
48 <u>EXE</u>	76.81	Monthly pmt

II. Graphics

The "COMP" mode must be used for the graph function. There are two applications of the graph function: Built In Function Graphs and User Generated Graphs.

The Built In Function Graphs are useful in illustrating the basic functions. There are 20 of these as follows: sin, cos, tan, $\sin^{-1}$, $\cos^{-1}$, $\tan^{-1}$, sinh, cosh, tanh, $\sinh^{-1}$, $\cosh^{-1}$, $\tanh^{-1}$, $\sqrt{}$, x^2, log, ln, 10^x, e^x, x^{-1}, $\sqrt[x]{}$. The graph for any of these can be produced by pressing "GRAPH, FUNCTION NAME, EXE." The graph may be erased by pressing "AC."

User Generated Graphs may be produced for any desired equation or combination of equations. However, the procedure here becomes more complex. Certain rules and procedures must be followed.

1. A range setting must be made. A typical range setting follows:

> Range
> x min : 0
>   max : 5
>   scl : 1
> y min : -5
>   max : 10
>   scl : 5

To set the range, press "RANGE." Range settings are made from the current position and proceed in the order: x min, x max, x scl, y min, y max, y scl. Input a numeric value at the current position and press "EXE." Use ⇨ and ⇩ keys to move to the next position to be set. Use (−) key for negative values. Press "RANGE" to exit. It is recommended that the range figures be slightly larger than the coefficients in the function being graphed.

2. Press "GRAPH" and enter the function. Press "EXE" and graph will be drawn. To erase a graph, press "SHIFT CLS EXE."

3. Two or more graphs can be overwritten, making it possible to determine intersection points.

4. The trace function: The coordinates for the intersection of two equations can be shown by using the trace function. Press "SHIFT TRACE." The pointer appears at the extreme left plot of the graph. The cursor key, ⇨, moves pointer to the right along graph. Each press of the cursor key moves the pointer one point, while holding down moves to the intersection of the graphs. The x-value appears. Press "SHIFT X—Y" and y-value appears. Figures shown are approximate, not exact. The trace function can also be used to find the x-values of a single equation.

EXAMPLE 5. Write a program to graph the following equations and evaluate the points of intersection:

$$Y = X^2 + 3X - 5$$
$$Y = 3X + 2$$

Calculator program:

PRESS	DISPLAY	COMMENT
MODE +		
GRAPH		
ALPHA X x^2		
+ 3 x ALPHA X		
− 5	X x^2 + 3 x X − 5	
EXE		Graph appears
GRAPH		
3 x ALPHA X + 2	3x X + 2	
EXE		Graph appears

<u>SHIFT</u> <u>TRACE</u>

⇨hold down	− 3.94	1st X-value
<u>SHIFT</u> X—Y	− 9.84	1st Y-value
⇨hold down	11.92	2nd Y-value
<u>SHIFT</u> X—Y	3.5	2nd X-value

Range and graph for Example 5:

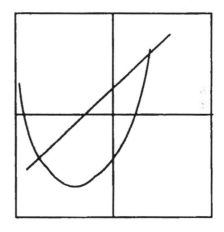

Range
x min : −5
   max : 6
    scl : 2
y min : −10
   max : 12
    scl : 2

EXAMPLE 6.   Write a program to show the point of tangency between the equations:

$$Y = X^3 + X^2 - 3X + 5$$
$$Y = 2X + 2$$

Range and graph for Example 6:

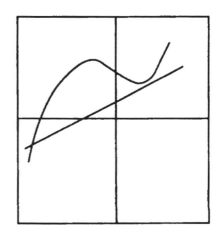

x min : −10
   max : 6
    scl : 1
y min : −10
   max : 15
    scl : 5

Calculator program:

PRESS	DISPLAY	COMMENT
MODE +		
GRAPH		
ALPHA X x' 3		
+ ALPHA X x²		
− 3 x ALPHA X + 5	X x' 3 + X	x² − 3 xX + 5
EXE		Graph appears
GRAPH		
2 x ALPHA X + 2	2 x X + 2	
EXE		Graph appears
SHIFT TRACE		
⇨Advance to tangency	0.96	X-value
SHIFT X—Y	3.93	Y-value

EXAMPLE 7. Write a program to show the x-values for the equation:

$$Y = X^2 - 20 X - 100$$

Calculator program:

PRESS	DISPLAY	COMMENT
MODE +		
GRAPH		
ALPHA X x² − 20 × ALPHA X		
− 100	X x² − 20 X − 100	
EXE		Graph appears
SHIFT TRACE		
⇨Hold down	− 4.36	1st value
⇨Hold down	24.36	2nd value

Range and graph for Example 7:

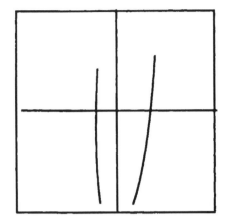

Range
 x min : -5
  max : 25
  scl : 2
 y min : -20
  max : 10
  scl : 0.5

EXAMPLE 8. Write a program to show the three x-values for the polynomial:

$$Y = X^4 - 7X^3 + 12X^2 + 4X - 16$$

Calculator program:

PRESS	DISPLAY	COMMENT
<u>MODE</u> +		
<u>GRAPH</u>		
<u>ALPHA</u> X x^y 4 $-$ 7 x		
<u>ALPHA</u> X x^y 3 + 12 x		
<u>ALPHA</u> X x^2 + 4 x		
<u>ALPHA</u> X $-$ 16	X x^y 4 $-$ 7 X x^y 3 $-$ 12X	x^2 + 4X $-$ 16
<u>EXE</u>		Graph appears
⇨Hold down	-1.17	1st value
⇨Hold down	2.18	2nd value
⇨Hold down	4.14	3rd value

Range and graph for Example 8:

Range
  x min : −5
    max : 10
     scl : 1
  y min : −10
    max : 15
     scl : 5

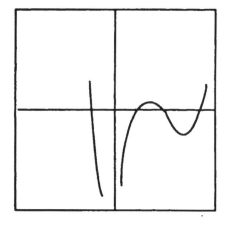

In this chapter we have explained the use of the FX-7000GA for formula programming and have illustrated its use for visualizing graphics. It is often helpful to illustrate the form of the graph for an equation.

13 · Programming and Special Techniques for the HP-32SII

The HP-32SII has a number of unique features. The topics which will be covered in this chapter are:

Fractions
Programming in the EQUATION mode
Use of the SOLVE function
Simple programming in the PROGRAMMING mode
Programming techniques
Numerical integration.

I. Fractions

The HP-32SII allows one to type in and display fractions, and to perform math operations on them. To enter fractions:

1. Key in the integer part and press ⌄ .
2. Key in the fraction numerator and press ⌄ again.

3. Key.in the denominator, then press ENTER to terminate digit entry. This also converts the fraction to a decimal.

If the number being entered has no integer part, type a ⌐ before the numerator. To turn the fractional display off or to change a decimal display to a fractional display, press "⌐ FDISP." One may perform any operation with fractions or between fractions and decimal numbers.

Entering Fractions

One can type any number as a fraction, including an improper fraction with the following exceptions:

1. The integer and numerator must not contain more than 12 digits total.
2. The denominator must not contain more than 4 digits.

Accuracy Indicators

The accuracy of a displayed fraction is indicated by the $\triangle$ and ∇ annunciators at the top of the screen.

1. If no indicator is lit, the fractional part of the internal 12 digit value exactly matches the displayed fraction.
2. If $\triangle$ is lit, the fractional part of the internal 12 digit value is slightly greater than the displayed fraction.
3. If ∇ is lit, the fractional part of the internal 12 digit value is slightly less than the displayed value.

EXAMPLE 1. Multiply 2 1/2 by 4 3/8.

PRESS	DISPLAY	COMMENT
2 . 1 . 2	2 1/2	
ENTER	2.5	
4 . 3 . 8	4 3/8	
x	10 15/16	
⌐ FDISP	10.9375	Fraction is exact

EXAMPLE 2. Multiply 1 2/3 by 3 1/4.

PRESS	DISPLAY	COMMENT
1 . 2 . 3	1 2/3	
ENTER	1.6667	
3 . 1 . 4	3 1/4	
x	5.4167	
⌐ FDISP	5 2/5 △	Decimal > fraction

EXAMPLE 3. Multiply 3 7/16 by 5 15/32.

PRESS	DISPLAY	COMMENT
3 . 7 . 16	3 7/16	
ENTER	3.4375	
5 . 15 . 32	5 15/32	
x	18 4/5	
⌐ FDISP	18.7988 ▽	Decimal < fraction

Many calculations involving fractions are simplified by using the fractional feature of this calculator.

EXAMPLE 4. Stocks are priced by fractions of a point. What is the value of 350 shares of a stock priced at 41 1/8 per share?

PRESS	DISPLAY
41 1 . 8	41 1/8
ENTER	41.125
350 x	14,393 3/4
⌐ FDISP	14,393.75

II. Programming in the Equation Mode

The HP-32SII provides two modes for programming, the EQUATION mode and the PROGRAMMING mode. The EQUATION mode is more suitable for short equations. In this mode the equation is written in algebraic format. The equation appears in the display. It is not assigned a label. To enter the EQUATION mode, press " ⌐EQN." To enter a symbol such as A, press "RCL A." The rest of the procedure is straight forward. One point should be noted—you cannot enter "x^2." It is necessary to use "y^x 2." When entry of the equation is complete, press "ENTER."

EXAMPLE 5. Write a program to determine the area of a circle with the equation A = $\pi d^2/4$.

PRESS	DISPLAY	COMMENT
⌐ EQN	Equation list top	Enter equation mode
RCL	■	Begins new equation
A ⌐ =	A =	
⌐ π		
x RCL D y^x 2		
÷ 4	A = π x D$^\wedge$ ÷ 4	
ENTER		Terminates equation

Solution for D = 5:

PRESS	DISPLAY
⌐ EQN ENTER	D = ?
5 RS	19.635

Solution for D = 25:

PRESS	DISPLAY
⌐ EQN ENTER	D = ?
25 RS	490.8739

Displaying and Selecting Equations

Equations that have been written are retained in the equation list. Press " ⌐EQN" and then up through the equations by pressing " ⌐ △" or down by pressing " ⌐ ▽."

To View a Long Equation

1. As the space in the display is limited to 12 characters, the "→" annunciator indicates more characters to the right. The annunciator "0" over the "Σ +" means scrolling is turned on.
2. Press "Σ +" to scroll the equation one character at a time to the right. Press "√x̄" to scroll characters to the left.
3. Press " ⌐SCRL" to turn scrolling off and on. When scrolling is turned off, the left hand end of the equation is displayed and unshifted top row keys perform their labeled functions. You must turn off scrolling if you want to enter a new equation that starts with a top row function.

Editing and Clearing Equations

You can edit or clear an equation that you are typing. You can also edit or clear equations saved in the equation list.

    To edit an equation that you are typing:

1. Press '⇦' repeatedly until you delete the unwanted number or function. If you are typing a decimal number and the " − " digit entry cursor is on, '⇦' deletes only the rightmost character. If you delete all characters in the number, the calculator will switch back to the " ▌ " equation entry cursor. If this cursor is on, pressing "⇦" deletes the entire rightmost number or function.
2. Retype the rest of the equation.
3. Press "ENTER" to save the equation in the equation list.

    To edit a saved equation:

1. Display the desired equation.
2. Press '⇦' (once only) to start editing. The " ▌ " equation cursor appears at the end of the equation.
3. Use '⇦' to edit the equation as described above.
4. Press "ENTER" to save the equation in the equation list, replacing the previous version.

Clearing:

1. Display the desired equation.
2. Press " ⌐ CLEAR."

    Following are further examples of programs in this mode:

EXAMPLE 6. Write a program for the solution of the Pythagorean theorem: $C = \sqrt{A^2 + B^2}$

PRESS	DISPLAY
⌐ EQN	
RCL C ⌐ =	
√x̄ RCL A y^x 2	
+ RCL B y^x 2)	
ENTER	

Solution for A = 4, B = 3:

PRESS	DISPLAY
⌐ EQN ENTER	A = ?
4 RS	B = ?
3 RS	5

Solution for A = 56, B = 78

PRESS	DISPLAY
⌐ EQN ENTER	A = ?
56 RS	B = ?
78 RS	96.0208

EXAMPLE 7. Write a program for the solution of the equation
$$D = \sqrt{a^2 + b^2 - 2ab \cos C}$$

PRESS	DISPLAY
⌐ EQN	
RCL D ⌐ =	
√x̄ RCL A y^x 2 + RCL B y^x 2	D = SQRT (A^ + B^
− 2 x RCL A x RCL B	−2 × A × B
x COS RCL C	× COS C)

ENTER

Solution for A = 600, B = 40, C = 45:

PRESS	DISPLAY
┌ EQN ENTER	A = ?
600 RS	B = ?
40 RS	C = ?
45 RS	572.41

Solution for A = 50, B = 65, C = 35

PRESS	DISPLAY
┌ EQN ENTER	A = ?
50 RS	B = ?
65 RS	C = ?
35 RS	37.42

III. Use of the Solve Function

This function has two applications:

1. It may be used to find the value of any variable in an equation when the other variables are known.
2. It may be used for the solution of polynomials. Equations must be written in the EQUATION mode for the use of this function.

Solving an Equation

1. Press " ┌ EQN" to display the desired equation. If the equation is in the equation list, press " ┐ △" or " ┐ ▽" until it is brought up. If this is a new equation, you may type over the equation in the display without affecting the displayed equation.
2. Press "SOLVE" and then the key for the unknown variable. The calculator then prompts for the values of the known variables.
3. For each prompt, enter the desired value. If the displayed value is the one you want, type "RS." If you wish to use another value, type over the one in the display and then press "RS."

EXAMPLE 8. The equation for the distance traveled and time for an object in free fall is:

$s = 1/2\, gt^2$ where s is distance traveled, g is the acceleration of gravity, 9.8 m/sec², and t is the time in seconds.

PRESS	DISPLAY
┌▸ EQN	
RCL S ┌▸ =	
0.5 × 9.8	S = 0.5 × 9.8
x RCL T yˣ 2	x T^ 2
ENTER	

Solve for S when R = 4 sec:

┌▸ SOLVE	
S	T?
4 RS	78.4

Solve for T when S = 150 m:

┌▸ SOLVE T	S?
150 RS	5.53 sec.

EXAMPLE 9. Using the equation $a = \pi d^2/4$ from Example 5, solve for the diameter of a circle having an area of 10 in.

PRESS	DISPLAY
┌▸ EQN	
┌▸ SOLVE	
D	A?
10 RS	3.5682 in

EXAMPLE 10. A very useful application of the SOLVE function is in the solution of the payment equation:

$$P = V\left[\frac{I/12}{1 - (1 + (I/12))^{-N}}\right]$$

where

P = the monthly payment
I = annual interest rate
N = number of monthly payments

PRESS	DISPLAY
⌐ EQN	
RCL P ⌐ =	P =
⌐ (RCL I ÷ 12 ⌐) ÷	(I ÷ 12) ÷
⌐ (1 − ⌐ (1 + ⌐ (RCL I ÷ 12 ⌐)	(1 − (1 + (I ÷ 12)
⌐) yˣ +/− RCL N ⌐)	y^ − N)
x RCL V	x V
ENTER	

Solve for P when V = 4000, I = 0.115, N = 48

⌐ SOLVE P	I?
0.115 RS	N?
48 RS	V?
4000 RS	104.36

Solve for V when P = 150

⌐ SOLVE V	P?
150 RS	I?
0.115 RS	N?
48 RS	5749.54

Solving a Polynomial

The SOLVE function may be used to find the roots of a polynomial. The calculator uses an iteration procedure for this operation. One must make a guess that the value of X falls between two points. The calculator either determines the root or displays "No root found." The process is repeated for other values. A polynomial may have several roots. The equation is entered as before. The solution procedure begins with a statement of the limits of the guess. For limits of 0 and 5 the statement would be 0 STO X 5.

EXAMPLE 11. Solve for the roots of the equation:

$x^3 + 2x^2 - 7x + 5 = 0$

PRESS	DISPLAY
┏ EQN	
RCL X y^x 3	X^ 3
+ 2 x RCL X y^x 2	+ 2x X^ 2
− 7 x RCL X + 5 = 0	−7 x X + 5 = 0

ENTER C

Solution

Try limits of 0 and 5.

0 STO X 5

┏ EQN

┏ SOLVE X   SOLVING

        No root found

Try limits of −5 and 0.

5 +/− STO X 0

┏ EQN

┏ SOLVE X   SOLVING

        −4.039

EXAMPLE 12. Solve for the roots of the equation:

$X^4 - 7X^3 + 12X^2 + 4X - 16 = 0$

PRESS	DISPLAY
Γ EQN	
RCL X y^x 4	X^ 4
-7 x RCL X y^x 3	-7 x X^ 3
$+ 12$ x RCL X y^x 2	$+ 12$ x X^ 2
$+ 4$ x RCL X	$+ 4$ x X
$- 16 = 0$	$- 16 = 0$
ENTER C	

Solution:

5 +/− STO X 0

Γ EQN

Γ SOLVE X   X = −1

0 STO X 2

Γ EQN

Γ SOLVE X   X = 2

3 STO X 4

Γ EQN

Γ SOLVE   X = 4

Interrupting the SOLVE Calculation

When the calculator uses a great deal of time, one may wish to halt the calculation and examine the progress. Use "C" or "RS" to halt the calculation. Use " Γ VIEW" to view it without disturbing the stack.

IV. Simple Programming in the Programming Mode

In this mode Reverse Polish Notation is used and programs are assigned a label. A comparison of the strengths of the two modes follows. RPN uses less memory and is a little faster. The EQUATION MODE is easier to write and read and automatically prompts for input of data.

Some of the same examples are presented in this section that were presented in the section on the EQUATION mode. This will enable the readers to make their own choices regarding their preference.

The programming capability of this calculator is limited by the amount of memory available which is 390 bytes. Numbers use 9.5 bytes each except for integer numbers from 0 to 99 which use 1.5 bytes each. All other instructions use 1.5 bytes each.

Users should also familiarize themselves with the clearing procedures which apply. "⌐ CLEAR" gives options for clearing: (X), (VARS), (ALL), and (Σ). During program entry the menu includes (PGM) which erases all program memory. As with other menu displays, the user should press the key in the top row directly under the selection.

There are two slightly different procedures for programming: single variable programs and multiple variable programs. We will first cover a single variable program. "⌐ PRGM" enters the programming mode. Next press "⌐ LBL" to assign a letter for labeling the program. Enter the program and press " ⌐ RTN" and "C C" to end the program.

EXAMPLE 13. Write a program for finding the area of a circle using the formula:

$$A = \pi d^2/4$$

Calculator program

PRESS	DISPLAY	COMMENT
⌐ PRGM		Enters program mode
⌐ GTO . .	PRGM TOP	
⌐ LBL A	A01 LBL A	Lables program
⌐ x^2 ⌐ π	A02, A03	
x 4 ÷	A04, A05, A06	
⌐ RTN C C	A07	Ends program

To execute this program, press the number which is the diameter of the circle, then "XEQ" and A. The area of the circle appears in the display. The area of a 10 cm diameter circle is 78.54 sq. cm. As long as this program is in the memory, it is only necessary to use this procedure, even after the calculator has been turned off. Consider a circle of some other diameter, say 13.75 cm. Press 13.75 XEQ A and the answer, 148.489 appears.

Testing a Program

A program may be tested by stepwise execution:

1. Make sure that program entry is not active.
2. Press " ⌐GTO" (label) to set the pointer at the start of the program.
3. Press and hold " ⌐ ∇" to display the current line. When you release " ⌐ ∇" the current line is executed and the result is displayed. To move to the preceding line, press " ⌐ Δ." No execution occurs.
4. Repeat step 3 for the remaining lines.

An example of this procedure for the area of a circle program follows.

KEY	DISPLAY
10	10
⌐ GTO A	10.000
⌐ ∇ (hold)	A02 x
(release)	100
⌐ ∇ (hold)	A03
(release)	3.142
⌐ ∇ (hold)	A04 x
(release)	314.159
⌐ ∇ (hold)	A05 4
(release)	4
⌐ ∇ (hold)	A06 ÷
(release)	78.540
⌐ ∇ Hold	A07 RTN
(release)	78.540

Multiple Variable Programs

Where more than one variable is used, the " ⌐ INPUT" function is used in the program. It is usually convenient to use the letter occurring in the formula. In the execution of the program the letter identifying the variable appears followed by "?." The "R/S" key is then pressed and the next variable is then requested. In writing the program, the input statements are made first. The program proceeds with "RCL" to enter the variable. When the instructions have been completed, press "RTN" (lable) and "CC" to end the program.

EXAMPLE 14. The Pythagorean theorem is one of the most frequently used formulas in engineering calculations. Write a program to solve the Pythagorean theorem:

$$C = \sqrt{A^2 + B^2}$$

PRESS	DISPLAY
⌐ PRGM	PRGM TOP
⌐ GTO . .	
⌐ LBL P	P01
⌐ INPUT A	P02
⌐ INPUT B	P03
RCL A ⌐ x^2	P04, P05
RCL B ⌐ x^2 +	P06, P07, P08
$\sqrt{x}$	P09
⌐ RTN C C	P010 Ends program

Following is the execution of this program, using values of 3 and 4 for A and B:

PRESS	DISPLAY
XEQ P	A = ?
3 R/S	B = ?
4 R/S	5

A second example, using values of 356 and 412:

PRESS	DISPLAY
<u>XEQ</u> P	A = ?
356 <u>R/S</u>	B = ?
412 <u>R/S</u>	544.5

Stopping or Interrupting a Program

Pressing "R/S" during a program inserts a STOP instruction. This will halt a running program until you resume it by pressing "R/S."

If an error occurs in the course of a running program, program execution halts and an error message appears in the display. To see the line in the program containing the error causing instruction, press "⌐ PRGM." The program will have stopped at that point.

Editing a Program

You can modify a progrm in program memory by inserting and deleting program lines. Even if a program line requires only a minor change, you must delete the old line and insert a new one. To delete a program line:

1. Select the relevant program (⌐ GTO (label)), activate program entry (⌐ PRGM), and press " ⌐ ∇" or " ⌐ △" to indicate the program line to br changed. If you know the line number wanted, press "⌐ GTO . (label)."
2. Delete line to be changed by pressing '⟻" The pointer then moves to the preceding line.
3. Key in the new instruction. This replaces the one deleted.
4. Exit program entry by pressing "C" or " ⌐ PRGM." To insert a program line, locate and display the program line preceding the point of insertion. Key in the new instruction. Subsequent program lines are moved down and renumbered accordingly.

The Catalog of Programs

The catalog of programs is a list of all program labels with the number of bytes of memory used by each label and the lines associated with it.

Press "⌐ MEM (PRGM)" to display the catalog, and press "⌐∇" or "⌐△" to move within the list. This catalog may be used to:

1. Review labels and the memory used by each program.
2. Execute a labeled program. Press "XEQ" or "R/S" while the label is displayed.
3. Display a labeled program. Press "⌐ PRGM" while the label is displayed.
4. Delete a specific program. Press "⌐CLEAR" while the label is displayed.

We will now proceed with more examples.

EXAMPLE 15. Write a program for the application of the law of cosines:

$$c = \sqrt{a^2 + b^2 - 2ab \cos C}$$

Calculator program.

PRESS	DISPLAY
⌐ PRGM	PRGM TOP
⌐ GTO . .	
⌐ LBL C	01
⌐ INPUT A	02
⌐ INPUT B	03
⌐ INPUT C	04
RCL A x^2	05,06
RCL B x^2 +	07,08,09
2 ENTER	010,011
RCL A x RCL B x	012,013,014,015
RCL C COS x +/- +	016,017,018,019,020
√x	021
↱ RTN C C	022

Following is the execution of this program, using values of 600 and 40 for the sides and 45 degrees for the included angle:

PRESS	DISPLAY
XEQ C	A = ?
600 R/S	B = ?
40 R/S	C = ?
45 R/S	572.42

A second example, using values of 36.5, 72.3, and 36 Degrees:

PRESS	DISPLAY
XEQ C	A = ?
36.5 R/S	B = ?
72.3 R/S	C = ?
36 R/S	47.85

EXAMPLE 16. Write a program for determining the monthly payment on a mortgage, using the formula:

$$Pmt = V \left[\frac{I/12}{1 - (1 + I/12)^{-N}} \right]$$

where

V = value of mortgage
N = number of monthly payments
I = annual interest rate

Calculator program.

PRESS	DISPLAY
꒐ PRGM	
꒐ GTO . .	
꒐ LBL M	01
꒐ INPUT V	02

⌐ INPUT N	03
⌐ INPUT I	04
RCL I 12 ± STO A	05,06,07,08
1 RCL A ± ENTER	09,010,011,012
RCL N +/− y^x +/−	013,014,015,016
1 ± $\frac{1}{x}$ RCL A x	017,018,019,020,021
RCL V x	022,023
⌐ RTN C C	024

Following is the execution of this program using the following values:

V = 25,000

N = 360

I = 0.09

PRESS	DISPLAY
XEQ M	V = ?
25,000 R/S	N = ?
360 R/S	I = ?
0.09 R/S	201.16

A second example using the following values:

V = 5000

N = 48

I = 0.115

PRESS	DISPLAY
XEQ M	V = ?
5000 R/S	N = ?
48 R/S	I = ?
0.115 R/S	130.45

V. Programming Techniques

This topic covers some of the more sophisticated techniques of programming:

1. Using loops with counters to execute a set of instructions of a number of times.
2. Using conditional instructions.
3. Using subroutines.

Loops

A loop is a program that repeats execution of a set of instructions. A good example is the determination of factorials. A factorial is the product of a series of consecutive integers beginning with 1. In the looping procedure the multiplication of the first two integers is performed and then the program is directed to repeat after adding 1 to one of the integers. The direction to repeat is performed by the "GTO" instruction. However, this calculator cannot go to a line number with this instruction. It must refer to a label.

EXAMPLE 17. This program illustrates the performance of a loop for determining factorials. A pause is inserted at two positions in the program to display the number and its factorial. This example is intended only to illustrate the function of a loop. The next example will illustrate the procedure for determining the factorial of a specific number.

Calculator program:

PRESS	STEP	COMMENT
⌐ PRGM		
⌐ GTO . .		
⌐ LBL F	F01	
0 STO A	F02,03	
1 STO B	F04,05	Initial storage
⌐ LBL D	D01	Iniate label D
1 STO A RCL A	D02,03,04	Add 1 each loop
⌐ PSE	D05	Read number

<u>STO</u> x B	D06	x previous factorial
<u>RCL</u> B	D07	
⌐ <u>PSE</u>	D08	Read factorial
⌐ <u>GTO</u> D	D09	Continue loop
⌐ <u>RTN</u> <u>C</u> <u>C</u>	D10	End program

To run program, press "XEQ F." Numbers and factorials appear:

1	1
2	2
3	6
4	24 and so on.

Conditional Instructions

There are many programs that involve making decisions. There are 12 comparisons available for programming.

Pressing "⌐ x?y" displays a menu for tests comparing x and y:

$\neq \; \leq \; < \; > \; \geq \; =$

Press the key in the top row immediately below the desired choice. Pressing "⌐ x?0" displays a menu for tests comparing x and 0.

EXAMPLE 18. This example illustrates the use of conditional branching to determine the factorial of a specific number. In this program the desired number is entered by "⌐ INPUT T." T will be in the Y-register. A, the number in the loop, will be in the X-register. The conditional test instructs the program to continue the loop as long as x < y. When the loop has been completed, the factorial is stored in memory B. The instruction "⌐ VIEW B" displays the answer.

Calculator program for n = 25:

PRESS	STEP	COMMENT
⌐ PRGM		
⌐ GTO . .		
⌐ LBL G	G01	
⌐ INPUT T	G02	Input n
0 STO A	G03,04	
1 STO B	G05,06	Initial storage
⌐ LBL H	H01	
1 STO A	H02,03	
RCL A	H04	
STO x B	H05	Loop
RCL T	H06	
RCL A	H07	
⌐ x?y (≤)	H08	Is step number < n ?
⌐ GTO H	H09	Return to loop
RCL B	H10	
⌐ VIEW B	H11	View factorial
⌐ RTN C C	H12	End program

Execution:

XEQ G	T?	
25 RS	1.551E25	

Subroutines

When a series of instructions is executed more than once in a program, it can be entered as a subroutine. A subroutine must be assigned a label.

It is applied in the program by "XEQ (label)." The instructions for a subroutine are entered after the main program. The subroutine must end with the instruction " ↱ RTN."

EXAMPLE 19. The application of subroutines will be illustrated in the following program for the solution of a quadratic equation:

$$x = \frac{-b + \sqrt{b^2 - 4ac}}{2a}$$

It can be seen that the solution provides two roots. A subroutine for the expression under the radical is applied twice in the program.

Calculator routine:

PRESS	STEP	COMMENT
↰ PRGM		
↰ GTO ⸳ ⸳		
↰ LBL Q	Q01	
↰ INPUT A	Q02	Data entry
↰ INPUT B	Q03	
↰ INPUT C	Q04	
RCL B	Q05	
+/−	Q06	
XEQ J	Q07	Call subroutine
+ 2	Q08, 09	
÷	Q10	
RCL A	Q11	Calculate and store root
÷ STO R	Q12, 13	
RCL B +/−	Q14, 15	
XEQ J	Q16	Call subroutine
− 2 ÷	Q17,18,19	
RCL A	Q20	Calcualte and store root

÷ STO S	Q21, 22	
RCL R	Q23	First root
⌐ VIEW R	Q24	
RCL S	Q25	Second root
⌐ VIEW S	Q26	
⌐ RTN C C	Q27	End main program

Subroutine J

⌐ LBL J	J01
RCL B ⌐ x²	J02,03
RCL A	J04
RCL C	J05
x 4	J06,07
x −	J08,09
√x	J10
⌐ RTN	J11

Execution for equation $x^2 - 20x - 100 = 0$:

PRESS	DISPLAY	
XEQ Q	A = ?	
1 RS	B = ?	
−20 RS	C = ?	
− 100 RS	24.142 First root	
RS	−4.142 Second root	

Execution for equation $3x^2 - 50x + 150 = 0$:

PRESS	DISPLAY
XEQ Q	A = ?
3 RS	B = ?
−50 RS	C = ?
150 RS	12.743
RS	3.924

VI. Numerical Integration

This calculator has a very good algorithm for performing numerical integration. Since the calculator cannot compute the value of an integral precisely, it approximates it. The display format setting affects the level of accuracy assumed for your function and used for the result. Integration is more precise but takes much longer when a large number of decimal places is specified. For the examples shown in this section we will set the display at "FX4."

To integrate an equation:

1. Key in the equation, using the " ┌⁺EQN" function. Press "ENTER" and "C." If the equation is not at the top of the equation list, scroll through the equation list by pressing " ⌐ △" or " ⌐ ▽" until the desired equation is displayed. Press "C."
2. Enter the limits of integration : key in the lower limit, press "ENTER," then key in the upper limit.
3. Press " ┌⁺ EQN."
4. Press " ┌⁺ ∫" and the variable of integration.

We will present four examples, ranging from the simple to the more complex.

EXAMPLE 20. Integrate the expression $\int_1^3 x^2\, dx$

Enter the equation:

PRESS	DISPLAY
⌐ CLEAR	
┌⁺ EQN	
RCL X y^x 2	X^2
ENTER C	

Integrate the function:

PRESS	DISPLAY
1 ENTER 3	
┌⁺ EQN	
┌⁺ ∫ X	$\int = 8.6667$

EXAMPLE 21. Integrate the function $\int_2^5 3\sqrt{4x - 3}\, dx$

Enter the equation:

PRESS	DISPLAY
Γ EQN	
3 x $\sqrt{x}$ Γ	3 x SQRT (
4 x RCL X − 3 Γ)	4 x X − 3)
ENTER C	

Integrate the function:

PRESS	DISPLAY
2 ENTER 5	
Γ EQN	
Γ $\int$ X	$\int$ = 29.4562

EXAMPLE 22. Integrate the function $\int_{-2}^3 (6 - x^2 + x)\, dx$

Enter the equation:

PRESS	DISPLAY
Γ EQN	
Γ (6 − RCL X y^x 2	(6 − X^ 2
+ RCL X Γ)	+ X)
ENTER C	

Integrate the function:

PRESS	DISPLAY
2 +/− ENTER 3	
Γ EQN	
Γ $\int$ X	$\int$ = 20.8333

EXAMPLE 23. Integrate the function $\int_0^2 \left(\frac{\sin X}{X}\right) dx$

As this function involves trigonometric functions, the calculator must be set in the radian mode.

Enter the equation:

PRESS	DISPLAY
¬ <u>MODES</u> (RD)	
Γ <u>EQN</u>	
<u>SIN</u> <u>RCL</u> X	SIN (X
Γ)	SIN (X)
<u>÷ RCL</u> X	÷ X
<u>ENTER</u> C	

Integrate the equation:

PRESS	DISPLAY
0 <u>ENTER</u> 2	
Γ <u>EQN</u>	
Γ ∫ X	∫ = 1.6054

If the calculator attempted to evaluate this function at x = 0, the lower limit of integration, an error (DIVIDE BY 0) would result. However, the integration algorithm normally does not evaluate functions at either limit of integration, unless the end points of the interval of integration are extremely close together or the number of sample points is extremely large. If these conditions are not present, the substitution of a very small number such as 0.0001 for the lower limit will allow the integration to proceed.

It is recommended that the user read the complete instructions on this topic in the user's manual as well as Appendix D to cover special conditions that might arise.

We have covered the main special functions that are available on the HP-32SII. As one can see, this is a very versatile instrument. Many people prefer the Hewlett-Packard calculator because of the convenience of Reverse Polish Notation.

References

Bassin, Milton G., Stanley M. Brodsky, and Harold Wolkloff (1969). *Statics and Strength of Materials*, McGraw-Hill, New York.

Calter, Paul (1973). *Problem Solving with Computers*, McGraw-Hill, New York.

Casio, Inc. (1989). *FX-7000GA Owner's Manual*, Japan.

Faires, Virgil Moring, and Robert McArdle Keown (1960). *Mechanisms*, McGraw-Hill, New York.

Harris, Norman C., and Edwin Hemmerling (1972). *Introductory Applied Physics*, McGraw-Hill, New York.

Hewlett-Packard (1978a). *Solving Problems with Your Hewlett-Packard Calculator*, Corvallis, OR.

Hewlett-Packard (1978b). *HP-33 Owner's Handbook and Programming Guide*, Corvallis, OR.

Hewlett-Packard (1990). *HP-32SII RPN Scientific Calculator Owner's Manual*, Corvallis, OR.

Johnson, Alfred, and Harry H. Chenoweth (1967). *Applied Strength of Materials*, McGraw-Hill, New York.

Malmstadt, H. V., G. C. Enke, and E. C. Torren (1963). *Electronics for Scientists*, W. A. Benjamin, Menlo Park, CA.

Oberg, Eric, and F. D. Jones (1964). *Machinery's Handbook*, The Industrial Press, New York.

Rickmers, Albert D., and Hollis N. Todd (1967). *Statistics: An Introduction*, McGraw-Hill, New York.

Sienko, M. J. (1967). *Chemistry Problems*, W. A. Benjamin, Menlo Park, CA.

Sorum, C. H. (1969). *How to Solve General Chemistry Problems*, Prentice-Hall, Englewood Cliffs, NJ.

Spotts, M. F. (1971). *Design of Machine Elements*, Prentice-Hall, Englewood Cliffs, NJ.

Texas Instruments, Inc. (1979a). *Owner's Manual, TI-55 Calculator*, Dallas, TX.

Texas Instruments, Inc. (1979b). *Calculator Decision Making Source Book*, Dallas, TX.

Texas Instruments, Inc. (1989). *Texas Instruments TI-68 Guidebook*, Dallas, TX.

Washington, Allan J. (1970). *Basic Technical Mathematics with Calculus*, Cummings, Menlo Park, CA.

Woods, Frederick S., and Frederick H. Bailey (1928). *Elementary Calculus*, Ginn, Lexington, MA.

Index

Printed and bound by CPI Group (UK) Ltd, Croydon, CR0 4YY

17/10/2024

01775692-0001